Education for the Industrial World

Education for the Industrial World

The Ecoles d'Arts et Métiers and the Rise of French Industrial Engineering

Charles R. Day

The MIT Press
Cambridge, Massachusetts
London, England

© 1987 Massachusetts Institute of Technology

All rights reserved. No part of this book may be reproduced in any form by any electronic or mechanical means (including photocopying, recording, or information storage and retrieval) without permission in writing from the publisher.

This book was set in Baskerville by Asco Trade Typesetting Ltd., Hong Kong, and printed and bound by The Murray Printing Company in the United States of America.

Library of Congress Cataloging-in-Publication Data

Day, C. R.
  Education for the industrial world.

  Bibliography: p.
  Includes index.
  1. Technical education—France—History—19th century.
2. Technical education—France—History—20th century.
3. Industrial engineering—France—History—19th century.
4. Industrial engineering—France—History—20th century.
I. Title
T121.D39   1987   607'.1044   86-27794
ISBN 0-262-04088-3

For Elaine

Contents

Acknowledgments

Over the past decade the Social Sciences and Humanities Research Council of Canada has financed much of the research that has led to this book, and I am very grateful for its generous support. I am obliged also to the Centre National de la Recherche Scientifique of France, the National Research Council of Canada, the Social Science Federation of Canada, and the President's Research Committee of Simon Fraser University for their assistance.

I have had the pleasure of discussing my work with many scholars and friends over the years. I am particularly obliged to Patrick Harrigan and John Weiss for their help and encouragement. I would also like to thank Linda Clark, Robert Fox, Wynn and Robert Gidney, André Grelon, Robert Koepke, Joseph Moody, Robert Palmer, Harry Paul, David Pinkney, Barrie Ratcliffe, Fritz Ringer, Terry Shinn, Peter Stearns, Mary Lynn Stewart, George Weisz, and Gordon Wright for many kindnesses.

The Société des Ingénieurs Arts et Métiers kindly placed their library and archives at my disposition and provided the photograph reproduced on the jacket of this book. I would also like to thank those who provided so much assistance in the preparation of the manuscript: Jenny Alexander, Bernice Henderson, Jeannie Kanakos, Susan Smith, Jocelyn Samson, Paul Horton, and Nelson Quiroga. My sons were patient and good humored with their father-author, and my wife, Elaine, to whom this book is dedicated, helped in ways too numerous to express.

Education for the Industrial World

1

Introduction

Quelle bizarrerie dans nos jugements! Nous exigeons qu'on s'occupe utilement et nous méprisons les hommes utiles.

Denis Diderot, *L'Encyclopédie*

Early in the nineteenth century France developed a highly centralized and rather rigid system of public education, which was established by Napoleon and called the *Université*. This system was designed mainly to meet the educational needs of the French bourgeoisie. The bourgeoisie was educated in the classical lycées and collèges, also founded by Napoleon, which taught them a general humanistic culture based on classical languages and prepared them for state examinations in the *baccalauréat*. The *baccalauréat* in turn opened the way to the universities and, for the best students, to the *grandes écoles*. The latter were elite professional schools providing privileged access to the upper civil service, the state engineering corps, and the liberal professions.[1]

The common people were educated in elementary schools, which were first organized by the law on public education of 1833 (the Guizot Law). These schools taught only the three R's, religion, and morality. Beyond the elementary level, higher primary schools and teacher training colleges provided two or three years of additional instruction for the sons of the petite bourgeoisie.

However, the industrial expansion of the nineteenth century saw the rise of new occupational groups of manufacturers, engineers, technicians, and skilled workers whose professional needs could not be met by either traditional primary or secondary education. These groups required intermediate technical and professional schools teaching the industrial sciences and other forms of useful knowledge. Although the public education system, the *Université*, was slow to recognize the need for new forms of technoscientific education, a network of technical and professional schools developed under other ministries and came to play an important role in advancing French industrialization and promoting avenues of social advancement for the common man. Despite their importance, and perhaps because they evolved outside the mainstream of French public education, these schools have been neglected by historians and their impact underestimated. This book traces the history of the intermediate technical schools in France, the most important of which were the Ecoles d'arts et métiers. During the past two centuries these schools have graduated more industrial technicians and engineers than any other school in France. We shall examine how the Ecoles d'arts et métiers cultivated the technical knowledge and skills that were of practical use in the workplace, how they shaped the identity and outlook of their students, and how they acted as stepping stones for the advancement of their graduates.

The French intermediate technical schools were to be found under several ministries. The Ministry of Industry and Commerce, established in 1831, oversaw a small but comprehensive set of schools, including by the end of the century the Ecoles nationales professionnelles and the Ecoles pratiques de commerce et d'industrie at the higher primary level, the Ecoles d'arts et métiers at the intermediate level, and the Ecole centrale des arts et manufactures in higher education. The Ministry of Public Instruction directed a network of higher primary schools (Ecoles primaires supérieures) established by the law on public education of 1833, a special modern program set up in the secondary schools during the 1860s, and engineering institutes established in the university science faculties around the turn of the century. In addition, ministries such as agriculture, fine arts, public works, and the army and navy possessed their own professional schools, as did the Catholic Church and other private associations.

Previous studies of French education have tended to focus on the *Université*, on the lycées and collèges, and on elite institutions such as the Ecole polytechnique, which provided the royal way to advancement in the state and military technical corps, and the Ecole centrale des arts et manufactures, which educated the sons of businessmen for executive engineering careers in industry.[2] Graduates of these schools considered themselves the technical elite of France, a view accepted by virtually all historians, sociologists, and other observers who assumed that the professional preeminence of the graduates of the *grandes écoles* necessarily gave them a preeminent role in industrialization. After all, Polytechnique and Centrale recruited from families of higher social status, and given the conservative nature of French education and society, one could only expect the schools to function as "mechanisms of reproduction" of the system of social stratification. Thus historians could easily dismiss the Ecoles d'arts et métiers and other less prestigious schools under the Commerce Ministry, such as the Ecoles nationales professionnelles and Ecoles pratiques de commerce et d'industrie, in the course of arguing that France faced a chronic shortage of skilled personnel and practically trained engineers. For the first century of their existence, the Ecoles d'arts et métiers were officially defined as training skilled workers and foremen, and even those observers most favorable to them (Charles Dupin, Arthur Morin, Henri Tresca) assumed that their graduates would rise only slightly higher than their fathers, becoming *ouvriers d'élite*, foremen and works supervisors.[3]

But in fact the graduates of the Ecoles d'arts et métiers and other intermediate technical schools succeeded in advancing themselves socially and professionally and made striking contributions to industrial technology. Their advancement came, without question, in a rigid and conservative social and educational system. But it was just that rigidity, which kept technical and vocational studies out of the *Université* for so many years, that obliged other ministries to establish specialized schools to meet changing technical needs. Though advancement for young men of modest origin was difficult, it was not impossible; the supposedly monolithic educational system possessed a number of cracks and fissures. Hence the theories of Pierre Bourdieu, Jean-Claude Passeron, Michel Crozier, Terry Shinn, and others[4]— that French educational institutions narrowly "reproduced" or perpetuated an inequitable class system by providing the "cultural capital" required by the bourgeoisie as a source of social legitimation —may be somewhat too narrow to fully explain the complex interrelationship of social mobility and industrial development in France.

In recent years several scholars have argued that French society was not as closed and her industrial technology not as backward as has been thought. In a study of the origins and early career choices of 27,000 students and graduates of 229 lycées and collèges during the 1860s, Patrick Harrigan found that the secondary schools attracted a surprising number of young men from the lower middle classes.[5] Peter Lundgreen, in his work on higher technical education in Germany, and Fritz Ringer, in a comparative study of France, Germany, England, and the United States, concluded that by the end of the nineteenth century the French system of engineering education was roughly comparable to the German, especially if one accounts for differences of population, size of industry, and resources.[6] Harry Paul has challenged the declinist thesis in science of Joseph Ben-David, Robert Gilpin, and Henry Guerlac, arguing that France was deficient in neither the sciences nor the applied sciences.[7] In the field of economic history François Crouzet, Maurice Lévy-Leboyer, and T. J. Markovitch have noted the steady, if uneven, growth of the French economy during the nineteenth century and the rapid development of the years prior to the Great War, and they have minimized the role of social factors in retarding economic development.[8] Finally, Charles Sabel and others have emphasized the role of medium industries, so common in France, in encouraging technical innovation, a point of importance to us because gadzarts were very active in such firms.[9]

Most of these studies have dealt with either higher or classical secondary education; in this work we shall concentrate on intermediate technical education in France. We hope to provide insight into various issues that have been debated by historians, sociologists, and educators in recent years: education and social mobility, collective career biographies, the relationship of education to industry, the formation of new technical and managerial groups, professionalization and credentialism, and the role of alumni associations as educational lobbies and professional organizations. We shall also examine the boarding school regime of the Ecoles d'arts et métiers as an instrument of social control and as an illustration of the Great Enclosure theory of Michel Foucault.[10]

Our study begins in chapter 2 with a discussion of the higher technical schools: the Ecole polytechnique, the Ecole centrale, the Conservatoire des arts et métiers, and the engineering institutes created in the universities around the turn of the century. In chapters 3 and 4 we shall turn to the development of intermediate technical and professional education, concentrating on the system of professional schools that evolved under the Ministry of Industry and Commerce. This system developed slowly over the nineteenth century until the turn of this century when, after a long jurisdictional dispute with the Ministry of Public Instruction, the Commerce Ministry obtained control over the Ecoles nationales professionnelles and the Ecoles pratiques de commerce et d'industrie. On the eve of World War I the number of schools under this ministry had grown considerably and, combined with related schools and institutions under other ministries, provided France with a solid network of technical schools.

Since the 1950s France has developed one of the most extensive systems of secondary technical schools in the world.

The remaining chapters deal with the Ecoles d'arts et métiers, the principal subject of this book. These schools are among the oldest technical schools in Europe, having celebrated their bicentennial in 1980. Founded experimentally by the Duke de La Rochefoucauld on his estate at Liancourt in the 1780s and officially as schools of Arts and Trades by Napoleon at Châlons and Angers in 1806 and 1811, they gradually grew in number with the creation of Aix-en-Provence in 1843, Cluny in 1891, Lille in 1900, Paris in 1912, and Bordeaux in 1964. In 1907 they became engineering colleges and in 1945 university-level schools of engineering. Since their founding they have graduated 62,000 engineers and technicians, more than any other school in France. By 1920 they were producing one fourth of all French engineers.[11]

In chapters 5 through 7 we provide a historical account of the Ecoles d'arts et métiers—their programs, organization, and goals. In chapters 8 and 9 we discuss life in the boarding school regime (the *internat*), the long struggle of the students against a harsh school administration and the accompanying hazing rituals that enforced strict solidarity among them. An analysis of student customs, traditions, and songs provides an insight into their collective psychology and how this affected their social relations and careers and the activities of the alumni association later on.

In chapter 10 we present the results of a large-scale intergenerational study of the social origins and careers (including the last position) of more than 2,000 graduates of the Ecoles d'arts et métiers, the gadzarts (gars des arts), between 1820 and 1910 (careers followed to the 1960s). We shall trace the social and professional advancement of graduates through the stages of industrialization and, in the process, evaluate their role in French industrial development. Finally, in chapter 11 we shall assess the contribution of individual gadzarts to various French industries. Our thesis is that gadzarts were far more successful in industry than their humble origins and the modest official goals of the schools suggested, and that in the process they made substantial contributions to the development of French industry and to industrial technology.

As technicians and mechanical engineers employed mainly in the mechanical and metallurgical industries, the gadzarts were in constant contact with the graduates of the Ecole centrale and (to a lesser extent) the Ecole polytechnique. The polytechniciens, trained for the top positions in the technical branches of the civil service and the military, were less likely than the centraliens and gadzarts to seek positions in industry during the nineteenth century. The centraliens were educated specifically to become industrial engineers at the executive level. The social and professional gaps between polytechniciens and centraliens, on the one hand, and the gadzarts on the other, illustrate the divisions that have characterized French engineering generally since the eighteenth century. Despite the performance orientation of their profession, engineers have been slower than doctors, lawyers, and architects to professionalize. One explanation for this goes back to the eighteenth century, when the fledgling profession was dominated by gentlemanly practitioners who were quite different from the self-made mechanics who had learned on the job. While the elite practitioners raised the prestige of the profession, in time their presence also contributed to the creation of a gap between theory and practice and between style and performance, when the strength of the profession actually lay in the synthesis of these qualities.

The gulf between the well-born engineers like the polytechniciens and trained craftsmen like the gadzarts had its parallel in other countries. In the United States, as Monte Calvert has shown in his study of mechanical engineers in the United States from 1830 to 1910, the well-educated heirs of the old gentlemanly shop culture in the tradition of Jefferson and Franklin looked askance at the low-born engineers produced on the job or by the secondary polytechnics and land-grant colleges.[12] But one can at least argue that in the United States there was a tradition of respect for practical work, which resembled in some ways the shop culture of the gadzarts. In England there was also a very strong tradition of respect for the educational value of practical experience. The practice of promoting able workers to engineering posts without benefit of formal education persisted, though English universities did begin to graduate certified engineers during the second half of the nineteenth century.[13]

In Germany the Fachschulen, Realschulen, and even the Technische Hochschulen had considerably less prestige than the secondary gymnasia and the universities, where pure science, mathematics, and classical languages were taught. According to Fritz Ringer, in 1879 German civil engineers protested against a proposal that graduates of nonclassical secondary schools be admitted to their branch of the profession after completing advanced studies. They were afraid that their standing would be lowered if the utility of classical studies in occupational practice was not affirmed.[14] By the 1890s, however, advocates of technical and scientific education had the ear of the Kaiser; the Technische Hochschulen were promoted to the university level and increasingly the applied sciences were taught in the universities. Thus by the beginning of this century engineering was advancing into higher education in Europe and North America, either in higher institutions of technology or in the universities themselves.

In France engineering has been a prestigious profession since the eighteenth century, associated with higher education and with the public services, notably with the state corps of roads and bridges, of mines, and with the military officers corps in artillery, fortifications, and supply. Each branch possessed its own preparatory school: the Ecole des ponts et chaussées (1747), the Ecole des mines (1783), various military schools, and the Ecole polytechnique (1794). The schools of roads and bridges, of mines, and the military schools were the "écoles d'application" for Polytechnique graduates who wished to specialize and thereafter move directly into top-level engineering posts in the great state corps. Later on the graduates of the Ecole polytechnique and the Ecole centrale also enjoyed privileged access to executive engineering positions in the semipublic railroads.[15]

Though polytechniciens showed little interest in careers in industry during the nineteenth century, they influenced industrial development through their control over the transportation system and natural resources and were particularly meddlesome to the contractors and entrepreneurs working on state projects. In the public services and railroads they prevented technicians like the gadzarts from being promoted to higher-level posts.[16] Worse, toward the end of the century polytechniciens began to "parachute" (*pantoufler*) into industry. This and the government's decision to revive the universities and establish engineering schools in the science faculties threatened to set back the Ecoles d'arts et métiers by a half century, reducing them once again to schools of arts and trades. The worst was avoided by the award of the engineering diploma in 1907, but the Ecoles d'arts et métiers were promoted only to the level of secondary engineering colleges just when most engineering schools were being placed in the university level. Not until after World War II did the Ecoles d'arts et métiers finally became institutions of higher education.

Despite the big social gap between them, a number of parallels existed between the Ecoles d'arts et métiers, the least prestigious of the technical schools, and the Ecole polytechnique, the most prestigious. Both were created under liberal auspices during the late eighteenth century and were converted into quasi-military establishments by

Napoleon. Students wore uniforms, were governed by the military code, and were confined in barrackslike boarding schools. Both schools, in their own way, were elite technical institutions training leaders of men, though of course at quite different levels. The Ecole polytechnique recruited the best technical minds from the upper bourgeoisie via the lycées and prepared them for high positions of leadership in the state services; the Ecoles d'arts et métiers recruited from among the most promising students of the more plebeian primary system and trained them for lesser managerial functions in industry. Both directed their graduates toward professional careers rather than toward research and development, which may help to explain France's increasing weakness in these fields vis à vis the Germans and later the Americans.

Neither Polytechnique nor the Arts et Métiers were part of the Université, the public education system under the Ministry of Public Instruction. The Ecole polytechnique existed under the Ministry of the Army, a separation that protected its "noble" status from the meddling of educational bureaucrats. The Ecoles d'arts et métiers were placed under the Ministry of Industry and Commerce, which directed a system of technical and professional schools generally considered to be of inferior status because of their utilitarian objectives. The Ecole polytechnique was the source of the social legitimation for the rising bourgeoisie; the Arts et Métiers provided a means of advancement for an industrial petite bourgeoisie of artisans, skilled workers, foremen, and small machine-shop owners. Each school produced its own esprit de corps and a distinctive kind of man. The "noble education" of the Ecole polytechnique represented the transformation of an older aristocratic style into an ostensibly meritocratic one; the "training" of young workers in the Ecoles d'arts et métiers produced a new type of industrial technician combining classroom instruction and practical experience.

The programs of the Ecoles d'arts et métiers were systematically practical; the gadzarts were never allowed to obtain more than a smattering of *culture générale*. They studied some French, no liberal arts, and little pure mathematics and science. Conversely, the programs of Polytechnique resolutely avoided anything practical. The students prepared in classical languages in the secondary schools and then studied theoretical mathematics and pure science rather than engineering and applied science. Though Polytechnique's programs were encyclopedic and frequently outdated, such recondite and nonutilitarian features symbolized their social and intellectual superiority. In France the goal of higher education was the reproduction of knowledge and the selection of elites rather than the advancement of science.[17]

Thus the polytechniciens received a far more elaborate education than they needed for the routine operation of the departments of roads, bridges, mines, telegraphs, and military artillery and engineering. In his study of the Ecole centrale, John Weiss concluded that the teaching of the industrial sciences during the nineteenth century was not inordinately difficult. For example, physical mechanics did not include quadratic equations until later in the century, and the Monge method of industrial drawing was fairly easy to teach. It is not surprising that the overeducated polytechniciens soon became bored with their tasks. In the roads and bridges department they frequently left the real engineering functions to their assistants, the *conducteurs*, some of whom were gadzarts. Weiss quoted one polytechnicien who stated that he had used his mathematics only twice in his career. The diploma of the Ecole polytechnique, Weiss concluded, was as much a social as an engineering credential.[18]

In his recent book on the Ecole polytechnique, Terry Shinn confirms that the school was a means of social legitimation for the bourgeoisie.[19] Because of its high entrance standards and difficult curriculum, it fostered the illusion of meritocracy while satisfying the need of the newly rich to establish a connection with the state, the traditional creator of elites in France. Once employed in the state corps, the polytechnicien never had to work hard or compete again; he could live the life of a leisured gentleman. While readily criticizing Polytechnique for its narrow recruitment, abstruse program, and wastage of fine minds, Shinn and other historians have found it difficult to believe that the gadzarts might have outperformed polytechniciens on the job. If the latter were overeducated, the gadzarts were just as surely undereducated—"mere technicians" and "glorified mechanics" in the estimation of most observers.[20] In this they echoed the view illustrated by the French saying: "The Polytechnicien builds a bridge and it collapses—he doesn't know why. The Gadzarts builds a bridge and it holds—he doesn't know why."

Michel Crozier stated that the power of technical experts can be either high or low, depending on whether they can control access to areas of crucial uncertainty.[21] Because of his leadership image and his "cultural capital" (in the words of Bourdieu), the polytechnicien frequently found himself in a position to exercise such control.[22] The gadzarts, on the other hand, was almost too productive, his performance too predictable. He was too easily pigeonholed into middle and upper-middle technical posts as plant managers and directors of planning—good positions but seldom the best ones. If he wanted to get to the top, he had to go either into a firm that favored practical men over fancy engineers (Cail, Fives-Lille, Renault) or into business for himself. If the polytechnicien needed a good technical specialist, however, he had only to hire a gadzarts.

Randall Collins argued that both education and technology are cultural resources whose use is shaped by their place in the organizational process, by the struggle of informal status groups for favorable power positions. In distinguishing between productive and political labor, Collins observed that "productive labor is responsible for the material production of wealth, but political labor sets the conditions under which the wealth is appropriated." He concluded that property in positions determines most of the social organization and status-group conflicts in everyday life. Under the Old Regime the wealthy bourgeoisie obtained aristocratic leisure by purchasing offices; in modern times "one invests in educational credentials."[23] The culture of specific groups is transformed into abstract credentials. For example, the French bourgeoisie transformed its traditional values of work, self-reliance, and thrift into competition for the diplomas of the Ecole polytechnique and other *grandes écoles*, which opened the way to safe, gentlemanly posts in the higher civil service. When they later began to parachute into private enterprise, the prestige that they had accumulated in the public services was such that they received similar preferment in industry, even though there was no legal monopoly involved and no guarantee that the polytechnicien possessed the actual qualifications for executive and technical duties. Thus the modern "sinecure society" came to embrace the private as well as the public sector, and "leisure has been incorporated into the job itself."

Collins further argued that the distinction between engineers, technicians, and mechanics has been exaggerated and that people acquire most of their professional skills through experience on the job.[24] He stated that engineering was quite adequately taught in secondary polytcchnics and that the move into the universities derived from social considerations rather than technical needs. The findings of this study generally tend to support this argument. Though sometimes lacking in flair and imagination, the gadzarts have proved to be very competent production engineers, industrial managers, and sometimes heads of companies. Industrialists and other observers have argued since the nineteenth century that the programs of the Ecoles d'arts et métiers, stressing the integration of classroom instruction and shopwork, were particularly appropriate for industry because of the constant interplay between conception and realization in the production process. Though the Ecoles d'arts et métiers have had their critics over the years, there have been few industrialists among them.

By the middle decades of the nineteenth century, a half century before they obtained the engineering credential, gadzarts were probably as capable as the centraliens of directing most aspects of the production process and were often more capable than the polytechniciens. Yet, in the language of Plato, the polytechniciens were men of gold, who needed only to be refined; the gadzarts, born in an age of iron, could at best hope to become men of brass, second-level managers and industrialists rather than true leaders of men. They were trained, not educated, licensed, not certified. Their *formation morale* did not involve polishing but rather the chiseling and molding of the rough parts in order to assure that they became reliable foremen and supervisors. Though the gadzarts understood the language of their bosses and how to translate their ideas into blueprints and drawings, they continued to speak the language of the workers on an everyday basis. Their tragedy was that, as middle-level managers and supervisors, they were neither one nor the other. They lived and worked in a world apart.

2

Higher Technical Education Since the Eighteenth Century

Ce n'est pas avec de belles paroles qu'on fait du sucre de betterave; ce n'est pas avec des alexandrines qu'on extrait la soude du sel marin.

François Arago

The engineering profession had been associated with the public and military services since the time of Louis XIV. During the eighteenth century engineers began to be trained in special state schools such as the Ecole des ponts et chaussées, the Ecole des mines, and the Ecole du génie de Mézières. The Ecoles d'arts et métiers and the Conservatoire des arts et métiers, founded during the Revolutionary and Napoleonic era, were also state schools and were the forerunners of popular professional education in France. The Ecole centrale des arts et manufactures, organized privately in 1829, was the first school of civil (that is, civilian) engineering. To adequately discuss the development of intermediate technical education in France, we must first define and discuss the higher schools.

We shall begin with an analysis of the Ecole centrale of Paris, a school with which the Ecoles d'arts et métiers had many ties. Both produced engineers and technicians for industry, mainly for the machine and metallurgical industries. We shall then turn to the Conservatoire des arts et métiers of Paris, a center of industrial research and a museum of techniques offering continuing education courses in the applied sciences. Placed under the authority of the Ministry of Industry and Commerce in 1832, it has shared many interests with the Ecoles d'arts et métiers. Next, we shall discuss the establishment of fifteen engineering institutes in the university science faculties in the late nineteenth century. These institutes competed with the Ecoles d'arts et métiers in recruiting students and in placing graduates in technical positions in industry after graduation. Finally, the promotion of the Ecoles d'arts et métiers to engineering schools, and later to *grandes écoles,* since World War II, invites a discussion of higher engineering education in France to the present day.

The Ecole Centrale des Arts et Manufactures

The Ecole centrale was established privately in 1829 to train high-level engineering executives for French industry. Founded by a group of prominent businessmen and scientists, the school was associated particularly with the railway, metallurgical, and machine-construction industries. Its costly three-year program (about 2,000 francs per year for tuition and room and board) appealed mainly to the industrial bourgeoisie. The Ecole centrale recruited fewer students from the liberal professions than did the Ecole polytechnique (10 versus 16 percent), from high officials and military officers (7 versus 15 percent), and from *propriétaires-rentiers* (20 versus 37 percent), but many more sons of businessmen and industrialists (35 versus 13 percent).[1] Students coming from laboring groups came to about 13 percent at Centrale as opposed to 5 percent at Polytechnique, many of whom had received scholarships or reduced tuition.[2]

In *The Making of Technological Man* John Weiss studied the Ecole centrale and its students in order to place the school in the broader context of the new engineering profession. Weiss said that the Ecole centrale "continued throughout the century to transform a social elite into a technological elite." It equipped the sons of the bourgeoisie with "the technological knowledge needed to transform wealth acquired through commerce, agriculture, the delivery of professional services, or manufacturing into streams of income stemming directly from the activities of industrialization." In short the school was forming a new type of technological man, the civil engineer, in an age of rapidly expanding industry.

The founders of the school disliked the abstract mathematical teaching of the Ecole polytechnique. One could define the "centraliens" (or "centraux," as they were called in the nineteenth century) by their dislike of "les messieurs X" and their privileges. They thought of themselves as returning to the scientific tradition of Gaspard Monge, that is, to a solid engineering education falling between the narrow empiricism of the British and the abstract teaching of Polytechnique.[3] The first two years of the school's three-year program provided a general scientific background based on four key subjects: physics, mechanics, chemistry, and descriptive geometry. The third year allowed for specialization. Although there were no workshops in the school, there were laboratories, and provision was made for practical applications. Auguste Perdonnet gave a course in railroad technology, and a steam engine was installed in the school in 1831. There were frequent visits and field trips to factories, industrial laboratories, and museums.

Unlike the Ecole polytechnique and the Ecoles d'arts et métiers, the Ecole centrale was a private school and was an *externat*; that is, the students were free to live where they wished in Paris. Discipline was the responsibility of professors and school officials and was not, as at Polytechnique and the Arts et Métiers, in the hands of military proctors perpetually at war with the students. In 1857 Centrale was taken over by the state and placed under the Ministry of Industry and Commerce. An alumni association was organized in 1862.

By the 1840s the school had produced a number of distinguished railway engineers, notably Auguste Perdonnet (a professor at the school), Jules Pétiet, and Camille Polonceau. In industry the school was favored by several leading industrial families: the Montgolfier dynasty in the paper and metallurgical industries; the Schlumberger and Dollfus families in the textile industry in Alsace; the de Wendels in metallurgy; and the Pereires in banking. René Panhard and Emile Levassor pioneered in automobiles and Louis Blériot in aviation; Gustave Eiffel demonstrated the uses of structural steel in designing and building the world's first skyscraper, the Eiffel tower. André Michelin transformed Clermont-Ferrand into a great tire and rubber center, and Armand Peugeot developed the Montbéliard region (Doubs) into an automobile capital. The centraliens were responsible for many inventions and new industrial processes.[4]

The centraliens came from the upper industrial classes; most were educated in classical lycées and thereafter received an advanced technical education. They should have produced significant thoughts about industry and the modern world, but as Weiss noted, despite some activity in the workshop movement in the revolution of 1848 and the sponsorship of the Société Philotechnique, the centraliens produced no Saint-Simon, nor any other major social thinker.[5] The school offered no courses in the liberal arts or business management and only later in modern foreign languages and industrial hygiene. The students carried a heavy load, almost entirely in science and engineering. Despite their classical secondary education, they conformed to the stereotype of the engineer: practical and hard-headed.

The centraliens prided themselves on having created the profession of civil (civilian) engineering in France and the first engineering society, the Société des Ingénieurs Civils, in 1848, thus ending the old association of engineering with the functionaries of the state and military services. They talked about "la science industrielle," the application of science to industry and to life, which was supposed to evolve into a new branch of knowledge and into a new technical culture; according to Weiss, however, this boiled down in practice to little more then a naive faith in the automaticity of progress and an ethic of order and work. Having failed to come to grips with the question of the social and economic consequences of industrialization, the centraliens finished by "leaving too much cultural territory to their opponents."[6] The polytechniciens, for all their defects, participated in the Saint-Simonian movement, created the Société Polytechnique to bring science to the working classes, and played an active role in the events of 1848 and in other social movements.

The Ecole centrale became public in 1857 and thereafter grew from about 300 students to 720 by the 1880s. By the end of the century its students were more bourgeois than those of Polytechnique. There were only 66 scholarship holders at Centrale from 1892 to 1895 as opposed to 248 at Polytechnique, and a number of these were gadzarts.[7] By the end of the nineteenth century, Centrale had produced almost 8,000 graduates, of whom 5,830 were living. Of the latter the great majority (5,044, or 87 percent) lived in France, and the rest (786, or 13 percent) resided in other countries or in the colonies.[8] A total of 4,727 were employed in France, of whom four fifths worked in industry and transport. Twenty-two percent worked in the mechanical industries, metallurgy, mining, and armaments, followed by 13 percent in the railroads. Beyond that, they were spread evenly over other industries: chemicals, food, textiles, utilities, banking, and so on. In all 85 percent worked in business, banking, and industry, the great majority within metropolitan France.

In table 2.1 we note the fairly close relationship of centraliens and gadzarts in terms of the industries in which they worked. Four fifths of the graduates from both schools were employed in industry and transport. While slightly more than one fifth of centraliens worked in the mechanical, metallurgical, and mining industries, one third of gadzarts did so. The centraliens were slightly more versatile; they went into chemistry and electricity and into public administration and teaching more readily. But generally the two groups were employed in the same fields—the centraliens more frequently as the bosses, the gadzarts as their production managers.

Table 2.1
Professions of graduates of the Ecole centrale des arts et manufactures and of the Ecoles d'arts et métiers by the 1890s

School	Ecole Centrale 1890s		Ecole d'Arts et Métiers, 1899	
	(n = 4727)	percent	(n = 3910)	percent
Industry				
Building, Public Works	357	8	297	8
Chemical	460	10	130	3
Engineers	548	11.5	495	13
Mechanical	375	8	828	21
Metallurgy, Mining, Arms	677	14	489	12
Textiles	134	3	115	3
Utilities	497	10.5	132	3
Misc. Industry	149	3	114	3
Agriculture	69	1	—	—
Business-Banking	172	4	112	3
Military	—	—	289	7
Public Administration	397	8	148	4
Teaching	130	3	63	2
Transport (R.R.s)	606	13	601	15
Miscellaneous	105	2	97	2
Unknown	51	1	—	—

NOTE: 80 percent of centraliens (3,803) and 82 percent of gadzarts (3,201) were employed in industry and the railroads.
SOURCES: Ministère du Commerce, de l'Industrie, des Postes et des Télégraphes, Direction de l'enseignement technique, *L'Enseignement technique en France*, I: *Etablissements nationaux et Ecoles supérieures de commerce* (Paris 1900), pp. 136–139. *Annuaire de la société des anciens élèves des écoles d'arts et métiers*, Liste des sociétaires par professions, 1900.

Table 2.2
Profession of fathers from selected schools, 1961

School	Business, Banking, Industry, Agriculture percent	Civil Service, Liberal/ Teaching Professions percent	Misc. and Unknown percent
Arts et Métiers	68	22	10
Centrale	52	38	10
Ecole Normale Supérieure	29	66	05
Polytechnique	41	55	04

SOURCE: F. Ringer, *Education and Society in Modern Europe* (Bloomington: Indiana University Press, 1979), p. 196.

In recent years the recruitment of the Ecole centrale, now located at Châtenay-Malabry near Paris, has changed. Fewer students are the sons of manufacturers, merchants, shopkeepers, and artisans, and a growing number are the offspring of salaried employees in government and private services. Thus in 1961 (table 2.2) slightly over half (52 percent) were the sons of industrialists, businessmen, bankers, and farmers, while 38 percent were the sons of civil servants, teachers, clerks, and professional people. Another 10 percent were miscellaneous artists, doctors, pastors, and rentiers. This caused Fritz Ringer to conclude that "Centrale must have changed course sometime between 1870 and 1960. Following the direction taken by Polytechnique, its older sister, it apparently began to develop ties to the technical civil service and to the public sector generally rather than confining its interests to private industry."[9]

Although a slim majority (52 percent) of the students at Centrale continued to come from business, industry, and agriculture, they were more likely to be salaried employees at various levels than individual owners. The Ecole nationale supérieure d'arts et métiers continues to recruit mainly from modest social groups, and yet even here the percentage of sons of small and medium employees in business and industry has risen greatly, while that of industrialists, small owners, and shopkeepers has declined. The Ecole centrale and the Ecole d'arts et métiers still draw from the productive sectors to a greater extent than do the Ecole polytechnique and the Ecole normale supérieure.

In terms of recent careers, in 1979 the alumni association of the Ecole centrale traced 197 of the 310 graduates of the class of 1969. The largest number worked in the mechanical, electronic, and telecommunications industries; the second largest in metallurgy, followed about equally by automobiles and petrochemicals; then came banking, public works and construction, aviation, and railways. The association also reported on the first jobs taken by 200 (of 300) graduates of the class of 1978. More than one third (36 percent) worked in the mechanical, electrical, and electronics industries, one fifth (22 percent) in petrochemicals, and one tenth each in building and public works and in engineering firms. Only 4 percent were employed in metallurgy and 3 percent in the railways.[10] The decline first of the railways and later of metallurgy indicates a general shift away from the traditional industries. The gadzarts have also largely abandoned the railroads and are going into metallurgy in decreasing numbers. Generally the new high-tech industries, so frequently located in sunnier climes and generating the excitement of recent breakthroughs, appeal to young engineers.

Today the Ecole centrale continues to provide a general engineering education, what the French call "la polyvalence technique." During the first two years of study the student takes courses in mathematics, science, the applied sciences, and engineering. During his third year the student may specialize in one or two of about twenty options. The rapid development of science and engineering in recent decades and the introduction of courses in the arts, foreign languages, and economics threaten to overload the curriculum, to make it encyclopedic and unwieldy. To avoid this the school attempts to teach the basic principles and concepts underlying the subject so that the individual can deepen his own knowledge during the third year of specialization and later through professional activity, documentary research, or further education.[11]

The Conservatoire National des Arts et Métiers (CNAM)

The origins of the Conservatoire lay in the collection of machines put together by Vaucanson from 1775 to 1782 during the reign of Louis XVI. In order to prevent their dispersion, Louis founded a "cabinet of the king's machines" and made them available to artisans and workers for inspection. The Revolution added various collections, including those of suppressed religious orders. Encouraged by the Abby Grégoire, in 1794 the Convention Assembly created a "depot of machines, models, tools, descriptions, and books of the various arts and crafts useful to public instruction." The purpose of establishing an industrial museum in the midst of revolution and war was to diffuse useful knowledge and awaken among French artisans and workers that spirit of invention in the mechanical arts that had given Britain such a lead in industrial technology.

Under the Empire, in 1804 the scientist Chaptal, minister of the Interior, organized a major industrial exhibition and associated the best French scientists with the Conservatoire, which by then possessed an excellent collection of 3,335 drawings and designs called "the portfolio of Vaucanson." Monge's new descriptive geometry was featured. According to Artz, Monge's system "represented exactly in drawings of two dimensions objects that exist in three, deducing from the exact description of objects all that follows necessarily from their forms and their respective positions." His discovery has been the root of all mechanical drawing and the graphic method used in engineering. The course in mechanical drawing taught in the Petite Ecole of the Conservatoire des arts et métiers from 1795 to 1874 was oriented toward the design and construction of machines and attracted a number of industrialists (such as Charles Dollfus and Eugène Schneider) as well as mechanics and skilled artisans. The professors, V. Le Blanc and later Jacques-Eugène Armengaud, a graduate of the Ecole d'arts et métiers of Châlons in 1829 (chapters 5 and 11), refined and popularized the techniques of mechanical drawing and published journals and books that contributed to the gradual emancipation of French industrial technology from dependence on the British.[12]

Under the Constitutional Monarchy, many of France's most distinguished scientists, including Arago, Berthollet, Dupin, Gay-Lussac, and Thénard, were associated with the Conservatoire.[13] In 1832 the Conservatoire and the Ecoles d'arts et métiers were placed under the authority of the Ministry of Industry and Commerce. By the 1840s professors offered public courses in geometry, descriptive geometry, mechanics, physics, chemistry, agriculture, and economics (J. B. Say). Aside from the Petite Ecole, however, the public courses were general and had limited relevance to industry. When Arthur Morin became director in 1849, he and his assistant, Henri Tresca, consulted more frequently with industrialists and engineers, creating a laboratory of industrial research and courses of a more practical bent. By the 1860s the Conservatoire attracted as many as 176,829 auditors, the highest ever.[14]

The generation after 1870 saw little progress, however, as the

Third Republic concentrated on the organization of compulsory primary instruction and a system of higher primary and intermediate technical schools. As a result the number of auditors fell to 63,000 in 1900. Then, Alexandre Millerand, minister of Industry and Commerce, introduced several new chairs and a laboratory of industrial research.[15] For the first time, course-work certificates and year-end exams measured students' proficiency. Although this made the Conservatoire rather more like a school, it continued to be quite open by French standards. There were no entrance requirements, and auditors were free to organize their programs as they saw fit.

In 1920 the Conservatoire, along with almost all of the schools under the Ministry of Industry and Commerce, was transferred to the Ministry of National Education under a semiautonomous technical education division. As a result of rapid technological change during the Great War, the CNAM introduced courses in aviation, internal combustion, and petrochemicals. The number of auditors began to grow again, reaching 141,000 in 1925. An engineering diploma was introduced in 1922 based on courses, laboratory and practical work, but it was so difficult to obtain that during the first 40 years only 1,329 were awarded.[16]

In 1952 two diplomas of higher studies, one technical and one in economics, were introduced. They were based on a cycle of three courses and two laboratories. To prepare for the engineering credential, it was necessary to have obtained one of these degrees. During the same year, the government authorized the establishment of regional centers at Lille, Mulhouse, and Lyon. By 1965, 59 such centers existed in cities all over France, with 25,000 auditors in the Paris region and 8,000 in regional centers. The 35 research laboratories included an Institut aérotechnique at Saint Cyr and research labs in electronics, radiology, and atomic physics.[17]

During the 1960s two thirds of the auditors were between 20 and 30 years of age. In the course on metallurgy in 1963, which had 8,000 registered students, 12 percent were workers, 5 percent highly skilled workers, 47 percent laboratory technicians, 11 percent other kinds of technicians, 14 percent draftsmen, and 9 percent employees. Most came from middle-level educational backgrounds, about half possessing the *baccalauréat* and most of the remainder possessing certificates and diplomas from technical secondary schools.[18] Workers having the apprenticeship certificate were insufficiently prepared for such advanced courses and were directed toward other centers of continuing studies.

In 1963 only 178 diplomas of advanced studies and 100 engineering degrees were awarded out of several thousand who sat the examinations. While allowing a small elite to emerge, the CNAM does not consider itself primarily a degree-granting institution. Its purpose is to offer courses and research facilities for those seeking to enrich their knowledge of industrial technology and in the process improve their prospects for advancement. Though students are free to organize their own time and set their own academic goals, elaborate information services assist them and maintain liaison with industry and other academic establishments. This is especially important in provincial cities where the regional centers rely heavily on the local universities and other academic institutions and research centers for their professors and facilities. CNAM is organized as a loose federation; the regional centers set up courses in accord with local economic needs, hire faculty, and maintain contacts with local industry and academic institutions, but they are subject to some supervision from the center to ensure a certain uniformity of standards and goals.

The University Science Faculties and Schools of Engineering

In most Western countries higher education means universities, but this is not necessarily the case in France. By the seventeenth and eighteenth centuries the universities were losing their vitality and were slowly being supplanted by state-created higher professional schools. In 1808 and 1811 Napoleon established the *Université de France*, which was not a school but rather a highly centralized system of academies and councils possessing a monopoly over secondary and higher education in France. In 1802 he founded a system of secondary schools called *lycées* and in 1808 set up the *baccalauréat*, a state certificate based mainly on classical languages and mathematics, which was passed successfully by about 5 percent of the secondary school students. The main purpose of the 32 university faculties of arts and sciences was to administer and grade the *baccalauréat* examinations and give public lectures to laypeople. Aside from the professional faculties of law, medicine, and theology, there were few students. The arts and science faculties had from 5 to 8 professors on the average, while law and medicine had 10 to 26.[19] The lycées had so clearly supplanted the universities that the more distinguished among them created post secondary preparatory sections for the *grandes écoles*. Most aspiring lycée professors studied for the *agrégation* in the Ecole normale supérieure rather than in the universities.

After the defeat of 1870, Germany emerged as the undisputed leader in scientific research and the application of science to industry, especially in chemistry and electricity. Under the Third Republic Jules Ferry, education minister and prime minister from 1879 to 1885, favored the reform of French universities in order to advance the sciences, build national strength, and form republican elites. Because there were so many vested interests in the education system, however, thorough reform of higher education proved impossible. Hence no single law created French universities as we know them today. University faculties in a given city were linked together loosely by a series of administrative decrees from 1885 to 1896. These new "universities" were granted a measure of financial autonomy, an increased number of scholarships, and, during the first years at least, more generous public financing. Aspiring secondary school professors were encouraged to take the *licence* in the universities. Such reforms led to rapid growth in enrollments, from 11,200 in 1876 to 42,000 in 1914.[20]

Of all the developments of these years, probably the most important was the appearance of applied science institutes in or alongside the science faculties. Scholars such as Harry Paul, Terry Shinn, and George Weisz have cited these engineering schools as the most significant development in higher education during the two decades prior to the Great War. These schools trained engineers and technicians and provided facilities for research, especially for local industries. Such activities were permitted because the normally rigid *Université* bent its rules to allow the applied science institutes (though not the faculties) to accept graduates of technical and professional schools who did not possess the *baccalauréat*, and they in turn were allowed to award degrees that did not confer the privileges of the state diplomas. This, and the government's insistence that the universities find local sources of support to supplement state credits, forced them to participate more fully in the economic life of their regions.

The success of the applied science institutes had several causes. The science faculties of Lille, Lyon, and Nancy had been successfully collaborating with local industries since the 1850s. Second, the growth of the chemical and electrical industries during the 1880s and 1890s, and the need to close the gap with Germany, contributed to the popularity of the institutes. But there were other inducements, especially the desire of the science faculties to attract more students. The faculties usually began by opening special industrial or agricultural courses and laboratories, financed by local authorities and government grants. The larger ones at Grenoble, Lille, Lyon, Nancy, and Toulouse, with more resources and closer ties to local industry, organized semiautonomous institutes, engaged in product testing and development, provided laboratory space for local industries, and sometimes got involved in advanced applied research.[21]

Nancy, the oldest, most active center of applied research, was composed of the Institut chimique, the Institut électrotechnique et de mécanique appliquée, the Institut aérodynamique, and the Ecole de Brasserie. The industrialists of Lorraine subsidized these institutions through the Société industrielle de l'Est and other associations. At Grenoble local industrialists formed the Société pour le développement de l'enseignement technique, which raised funds for the Institut polytechnique, the Institut électrotechnique, and an industrial testing service. By 1912 the ensemble of institutes and schools at Grenoble gave the faculty the third largest enrollment in France, with 468 students, of whom 356 attended the Institut polytechnique. Together Grenoble and Nancy produced more graduates than the other 13 faculties combined. In 1908 they accounted for 178 of 248 graduates, or 72 percent, and by 1928, 286 of 479, or 60 percent. Lyon, Lille, and Toulouse grew rapidly in the years before the war.[22]

In Paris the science faculty had plenty of students (1,300 in 1900), and there were already a number of municipal technical and professional schools in existence; nevertheless the municipality and private groups established the Ecole municipale de physique et de chimie industrielles in 1882, which had an excellent laboratory where Marie and Pierre Curie did much of their research. The Ecole supérieure d'électricité (1894) and the Ecole supérieure d'aéronautique (1909) were both established privately in Paris.[23]

Most of the applied science programs ran for three years and provided a solid engineering education. By 1913 the institutes were granting nearly 500 industrial diplomas, mostly in chemical and electrical engineering. This constituted about one fourth of France's annual output of school-trained engineers.[24] The *grandes écoles* of engineering (Polytechnique, Centrale, and so on) produced about 500 engineers for industry and the public services. The Ecoles d'arts et métiers furnished about the same number of mechanical engineers and technicians.

Harry Paul found that the social origins of the students varied with the nature of the program: "In the 1890s most of the students in applied chemistry in Paris were the sons of industrialists or of directors of manufacture in chemical industries, although a few sons of civil servants and of businessmen outside the chemical industry were also recruited."[25] The applied science institutes also recruited from higher primary professional schools such as the Ecole La Martinière of Lyon and the Ecoles Colbert, Diderot, Lavoisier, J. B. Say, and Turgot of Paris, which recruited from among the sons of small industrialists, skilled workers, and artisans. Indeed, they drew their students from more or less the same clientele as did the Ecoles d'arts et métiers, which is one reason why the latter feared their competition.[26]

On the other hand, the better graduates of the Arts et Métiers sometimes continued their studies in the more advanced institutes, especially in electrical engineering. They constituted about one third (150) of the students at the Institut électrotechnique of Grenoble in 1921, and they were the main supplier of students between 1894 and 1925 to the prestigious Ecole supérieure d'électricité of Paris (408 of 2,804 students). Paul Janet, director of the school and a foremost scientist in his own right, said that the gadzarts frequently outdistanced the graduates of Polytechnique, Centrale, and other *grandes écoles* at "Supélec."[27]

On the basis of his study of the engineering institutes and French science in general, Harry Paul challenged the declinist thesis put forth by Joseph Ben-David, Robert Gilpin, Henry Guerlac, and others.[28] They had argued that French scientific education produced a small, theoretically trained elite incapable of responding adequately to industrial growth. The failure to train sufficient skilled personnel helped explain France's slow industrialization. Critics of French science also stressed the weakness of French universities, the co-opting of outstanding scientists to administrative positions, the under funding of scientific research, and the confinement of genuine research to special institutions such as the Collège de France, the Muséum d'histoire naturelle, the Ecole pratique des hautes études, and the science section of the Ecole normale supérieure.

Without denying the validity of some of these points, Paul emphasized French achievements in the fields of medicine, physiology, geology, and in the applied sciences.[29] He minimized the issue of institutional barriers to research, stressing the flexibility of the *Université* in the generation before the Great War. My own findings on intermediate technical education coincide with Paul's and suggest that France trained sufficient technical personnel to meet industrial demand in the decades before the Great War.

Terry Shinn and George Weisz accepted the evidence provided by Harry Paul and the author of this work of a growing interest in technical education and industrial research in France during the decades before the Great War, but they argued that this may have diverted interest and resources from scientific research toward a narrow vocationalism. Shinn argued that this may even have contributed to the decline of French science from 1900 to 1914, a point that Weisz criticized for being based on insufficient evidence.[30] Weisz stated that French scientists in the nineteenth century were "probably more willing to serve industrial interests than their counterparts in German universities." Though uncertain as to whether this harmed basic research, he concluded that "applied research provided French industry with few striking technological advances." The French chemical industry, for instance, performed well during this period but was hardly at the forefront of technological innovation.[31]

Paul agreed that the German lead was not likely to be overcome by the creation of a few engineering schools, but he disagreed with both the thesis of "excessive vocationalism" and the argument that the applied science institutes produced "few striking technological advances." He argued that by 1910 the science faculties in Grenoble, Lille, Nancy, and Toulouse had become "significant centers of scientific activity because of direct links with regional industries."[32] As such they helped fill the gap caused by the absence of information on the latest scientific and technological innovations important for industry. Paul thus insisted on the close connection between "pure" and "applied" science, citing the work of Pasteur in bacteriology for agricultural industries (brewing, distilling) at Lille from 1854 to 1857; of Mahistre in the efficiency of steam engines and the resistance of materials; and of Auguste Lamy in the application of heat in industry.[33] In 1912 the faculties produced two Nobel prize winners in chemistry, Paul Sabatier at Toulouse and Victor Grignard at Nancy. Prize-winning scientists such as Cuénot, Wahl, Haller, Janet, and Friedel were all associated with the institutes.[34] Finally, the new engineering schools in the universities broadened what had been a rather narrow set of schools of industrial engineering, limited mainly to the Ecoles d'arts et métiers and the Ecole centrale.

The Universities Since the Great War

The long period of industrial stagnation in France, running roughly from 1930 to 1950, saw little progress in higher technical and engineering education. The hopeful developments described by Harry Paul in the science faculties failed to materialize. Neglected during the interwar period, the universities have become institutions of mass education since the 1950s. In the French system the *bachelier*, the possessor of the *baccalauréat*, may register in the university faculty of his choice without other formalities. At the beginning of the century few candidates managed to pass the difficult examinations for that diploma; today the figure stands at about 60 percent, so the faculties must accept a wide range of students. The best students attend the *classes préparatoires* of the lycées, university-level cramming courses for the difficult entrance examinations for the *grandes écoles*. A few universities (Grenoble, Lyon) have tried to introduce more exacting entrance requirements, but the Ministry of National Education in Paris always overrules their efforts in the name of uniformity and general access. Critics point out that the absence of entrance standards explains why half the university students never obtain a diploma at all, and they decry what they call "la sélection par l'échec."[35]

There are, then, three levels of engineering schools in France. At the summit are the traditional elite schools of Paris: Polytechnique, Mines, Ponts et Chaussées, and Centrale, plus the Ecole supérieure de physique et de chimie (1882), the Ecole supérieure d'électricité (1894), and the Ecole supérieure d'aéronautique (1909). At the top of the next echelon one finds the Ecole nationale supérieure d'arts et métiers (ENSAM), which legally became a *grande école* in 1974, followed by the Ecoles nationales supérieures d'ingénieurs. On a lesser level are the Instituts nationaux des sciences appliquées, introduced in 1957, and the 60 science faculties, which have less prestige than ENSAM and the Ecoles d'ingénieurs, with the exception of several of the Paris faculties, Grenoble, Strasbourg, and Toulouse.

Finally, a number of two-year technical schools, called the Instituts universitaires de technologie (IUTs), train technicians for industry. The number of industrial technicians in France has grown rapidly from 193,206 in 1954 to 758,690 in 1975 (293 percent). They are trained in the IUTs, the lycées techniques, the Conservatoire des arts et métiers, the Ecoles nationales professionnelles (assimilated to the lycées techniques in 1960), in continuing education programs, and privately (*autodidactes*). Technicians find it difficult to become engineers through promotion on the job, and only about 10 percent make it, although they do constitute most of the *ingénieurs maison* who have advanced through the ranks and who have averaged about one fourth of the industrial engineers in industry since the Second World War.[36]

The Grandes Ecoles Since 1945

What is a *grande école* and how is it recruited? For an engineering school to be officially recognized as such it must be certified by the *Commission des titres d'ingénieurs*. This commission was organized by the law of 10 July 1934, which gave legal protection to the engineering diploma, and is composed of representatives from interested ministries, from the schools, and from the professions. Today there are 143 certified schools of engineering under nine different ministries (Agriculture, Defense, Environment, Industry, Interior, Posts and Telecommunications, Health and Welfare, Transport, and Education). These constitute almost three fourths of the 200 *grandes écoles* in France. The number of engineers graduating annually from the technical *grandes écoles* and other specialized schools comes to 12,000, plus 3,000 from the universities, a total of 15,000. The total number of engineers in France now stands at about 250,000, up from 55,000 in 1920.[37] The 143 engineering schools accept more students than they once did, the average coming to about 680 (grand total of 97,000 students), but the main reason for the growth is that two fifths of all *grandes écoles* have been created since 1946:[38]

1747 to 1899: creation of 41 schools

1900 to 1918: creation of 23 schools

1918 to 1940: creation of 22 schools

1946 to 1978: creation of 57 schools

The *grande école* is distinguished above all by its recruitment. Whereas the universities must accept all *bacheliers* who present themselves, the *grandes écoles* recruit *sur concours*, that is, from a limited number of openings each year, which is usually set according to the number of openings available in the profession for which the school prepares. A minority of 226 elite lycées, one fourth of which are located in Paris and three fourths in major provincial cities, offer 1,200 classes to 42,000 students (32,000 in scientific and technical classes). These are called *classes préparatoires* (*prépas*). Their programs last one or two years, but since many redouble the second year, one can easily spend three years in the *prépas*. Two thirds of the candidates obtain entry into a *grande école*, often after more than one try.[39] The third who are not so lucky used to be left with nothing to show for their years of study, but they are now awarded a two-year university certificate and may be accepted into certain of the less prestigious higher schools, such as the Ecoles nationales supérieures d'ingénieurs, or into the university faculties.

The *classes préparatoires*, though given in the lycées, belong to higher education. The number of their students, aged 18 to 21, constitutes only 2.5 percent of that age group in France (42,000 of 1.7 million). Once admitted to a *grande école*, they will have another three years of intensive work, sometimes followed by further specialized study. Most of the incoming students at the Ecole supérieure d'électricité, for example, are already engineers. An increasing number take the three-year doctoral program in engineering.[40]

In response to criticism that the great majority of those admitted to the *grandes écoles* come from affluent families, these schools have broadened their recruitment to accommodate students from other sources. Today 75 percent come via the *concours*, 10 percent are admitted directly on the basis of their secondary school record and/or their high standing in the *baccalauréat* examination, and 15 percent arrive via the *passerelles*. The engineering schools take from 20 to 40 percent of their students from the *passerelles*.[41] These "bridges" bring students from several sources: students from the universities who have done two years of work and have obtained the diplôme universitaire d'études scientifiques (DUES); graduates of the two-year Instituts universitaires de technologie; certificate holders from the Conservatoire des arts et métiers and other continuing education programs; and the graduates of various vocational and professional schools.

The *grandes écoles* have always adapted to changing circumstances and public criticism; they have replied to the charge that they recruited exclusively from the upper classes by seeking the brightest students from all social groups. Behind their willingness to adapt to change lies one of the best-organized educational lobbies in the world. Once the Faure reforms of higher education were implemented in response to the events of 1968 and the way appeared open to the democratization of higher education, defenders of the *grandes écoles* organized the Conférence des grandes écoles in 1969 and the Comité national pour le développement des grandes écoles (CNGE) in 1970. The Conférence des grandes écoles directly represents the elite schools, coordinating their activities and sponsoring conferences for discussion and exchange of ideas. It also maintains liaison with the CNGE, the representatives of the *classes préparatoires* of the lycées, and the various professional associations that defend the schools' interests. The CNGE is composed of the directors of the schools, representatives from the professions (especially from business and industry), and the alumni associations. It concerns itself with lobbying, public relations, and the defense of the schools' key interests: their *classes préparatoires*, their privileged access to the professions, budgetary allocations, entrance standards, and programs and diplomas.[42]

The Savary Law

The election of a socialist government in 1981 under François Mitterand and the reform of higher education presented by the education minister, Alain Savary, threatened the privileges of the *grandes écoles* and led to furious resistance on their part. Savary wanted to end the fragmentation of higher education by assimilating the *classes préparatoires* of the lycées to the universities and by unifying all branches of higher education under the Ministry of National Education, including the *grandes écoles techniques* scattered under nine different ministries. Of the schools affected, about half (the Ecole nationale supérieure d'arts et métiers, the Ecole centrale, the Ecoles nationales d'ingénieurs, the Instituts nationaux des sciences appliquées, and the university faculties), including about three fourths of the students, were already under the jurisdiction of the Ministry of National Education. But many of the schools located under other ministries were among the most prestigious (Aéronautique, Polytechnique, Mines, and Ponts et Chaussées, for example), and they were far from happy about the prospect of assimilation to the *Université*.[43]

They were particularly exercised about the threat to the *cours préparatoires*. Indeed any threat to the "prépas" is considered a threat to the entire *grande école* system, for through these classes the latter monopolize the best students. For example, in 1983 the technical *grandes écoles* accepted 11,000 of 32,000 applicants from the preparatory classes. The lucky ones receive the most thorough education that France has to offer; there is, for example, one professor for 11 students in the "prépas," a much better percentage than in the universities. So it is not surprising that a powerful lobby against the Savary Law was organized by the Comité national pour le développement des grandes écoles, the alumni societies, and various professional associations, including the Conseil national des ingénieurs français, representing 250,000 engineers and 50,000 higher technicians. These federations have very close ties to business and industry and to the higher civil service and represent a virtually insurmountable force protecting the monopolies and privileges of the *grandes écoles*. Their spokesmen argued that rigorous selection, if done without regard to class or origin, is preferable from the point of view of the working classes to the random admission procedures of the universities, "la sélection par l'échec." Savary's proposed law, they said, would destroy the "prépas" in the lycées, a system that had proved itself, in favor of a new and inefficient system that only postponed the selection process.[44]

Second, these groups dislike the prospect of the *grandes écoles* being assimilated to the universities. Business and industrial groups have never been fond of the universities. During the nineteenth century they associated them with the bureaucratic mandarins of the public education system, the *Université*. More recently, and especially since the *événements* of 1968, they consider them shot through with radicalism and permissiveness, producing unreliable employees. They prefer the better-trained and more tractable graduates of the special schools, whose orientation is systematically professional.

Although the Savary Law passed in the autumn of 1983, by then most of its essential features had been cut or seriously modified. Originally article 9 of the law gave the government the power to transfer schools by decree to the Ministry of National Education. The weaker amendment that passed stated that before any such transfer can take place, the administrative council of the school in question and the minister responsible for it must give their consent, which is highly unlikely. The best that Savary could come up with was an interministerial commission (Commission interministérielle de perspective et d'orientation de formation supérieure, in the quaint jargon of the rue de Grenelle), the nature and composition of which are to be decreed at a later date. The *cours préparatoires* remain unaltered. The law does limit the representation on governing boards and school councils of the alumni and professional and industrial associations.[45]

Though higher education in France has grown very rapidly since World War II, it retains its hierarchical and segmented character. Secondary education has grown even more rapidly and has been far more fundamentally reformed since the war. In the following two chapters we shall examine the development of intermediate technical and modern instruction since the nineteenth century in the context of secondary education in France.

3

Intermediate Professional Education, 1815–1914

Mécanique: adjectif désignant les professions qui semblent demander plus d'exercice aux bras qu'à l'intelligence.
Dictionnaire de Conversation, 1865 edition

The industrial expansion of the nineteenth century saw the rise of new occupational groups of manufacturers, managers, technicians, and skilled workers whose professional needs could not be met by traditional primary or secondary education. The primary schools taught the three R's to the average man and were terminal in nature; the secondary schools taught classical languages and abstract mathematics to the bourgeoisie, preparing them for the liberal professions and the higher civil service. Neither system, however, suited the professional needs of the new business and industrial classes, who required schools emphasizing applied science, modern languages, economics, and useful knowledge. After 1830 French governments saw the need for intermediate professional education, *l'enseignement professionnel*, but could not decide whether this program should be pitched at the higher primary level, appealing to skilled workers and farmers and including manual exercises, as industrialists wished, or whether it should be secondary and general, suitable for the new middle classes, as most education officials (*universitaires*) preferred.

The Guizot Law and the Higher Primary Schools

The debate over whether to place professional education in a new system of higher primary schools or into secondary modern programs in the lycées and collèges continued for half a century, from the July Monarchy into the Third Republic. The law on public education of 1833 (the Guizot Law) opted for the higher primary solution. This law obliged municipalities to maintain elementary schools for boys, the departments to open teacher-training colleges, and cities and towns of over 5,000 inhabitants to establish higher primary schools (écoles primaires supérieures). The higher primary schools were to prepare the petite bourgeoisie for lesser supervisory positions in commerce and industry.[1]

In France, however, this social group disdained the inferior primary system, preferring that their sons study Latin for two or three years in a lycée because of the prestige that such studies conferred. Further, many industrialists, workers, and artisans favored apprenticeship and learning on the job to postelementary instruction. And so by the 1840s most of the higher primary schools had failed. The education minister, Salvandy, attempted to organize a professional program in the secondary schools in 1847, especially in the local municipal colleges (collèges communaux), but this experiment was swept away in the Revolution of 1848.

During the reaction that followed the Revolution of 1848, conservatives argued that the instruction of the common people beyond the three R's produced *déclassés* and *demi-savants* who abandoned productive work for clerkships in various bureaucracies and who, as intellectual failures, were particularly susceptible to subversive ideas.[2] Consequently, the Falloux Law of 1850 modified the Guizot Law by making it easier for private organizations (notably the Church) to open schools, cutting back on the programs of the teacher-training schools, and generally seeking to widen the gap between primary and secondary education. The law did not mention either professional education or the higher primary schools. Many of these simply disappeared, though some professional programs survived in the municipal colleges, and a few of the higher primary schools located in industrial cities continued as municipal professional schools.

Professional, Special, and Modern Education

Under the Second Empire (1852–1870) Napoleon III instructed his education minister, Hippolyte Fortoul (1851–1856), to increase the science content in secondary education and encourage the creation of local professional schools in industrial cities. Fortoul responded by introducing a science program in the lycées alongside classical studies. This *bifurcation* evoked furious resistance from *universitaires* and other critics as a threat to *culture générale* and was later abolished by Victor Duruy. Fortoul also subsidized several new professional schools at Lille, Menars, Mulhouse, Rouen, and Toulon but for reasons not entirely clear soon lost interest in them.[3] His successor, Gustave Rouland (1856–1863), rescued the Institut industriel de Lille and the Ecole professionnelle de Rouen, which continued as schools for the sons of businessmen and industrialists, but the other three schools, training skilled workers, failed.

By 1860 no solution had been found to the question of professional instruction, although the Free Trade Treaty with England of 1860 and the Expositions in Paris (1855) and London (1862) brought home to the French the need to compete with their rapidly advancing neighbors. This led to two important government investigations into the question of industrial education in France, one sponsored by the Ministry of Industry and Commerce in 1863–1864 (of which more later) and the other by Rouland in 1862. Rouland's investigation revealed that one sixth of the students in the lycées and almost half of those in the municipal colleges were already enrolled in various professional courses that had developed, more or less on their own, during the preceding two decades. However, the secondary school inspectors reported that most of the programs were poorly organized and lacking in professors trained in the applied sciences and that the students, mainly the sons of artisans, farmers, and shopkeepers, were frequently ill-prepared for secondary studies. Many were simply there to acquire some "polish" (*dégrossir*) in order to raise the family's standing in the neighborhood or village. One third of the students attended modern programs in Catholic schools, which was one of the main reasons why public education officials decided to organize such a program under state auspices.

In 1862 Rouland presented a bill on l'enseignement professionnel to Parliament, but in mid-passage he was replaced by Victor Duruy (1863–1869), who spent another year and a half revising the bill, which passed in 1865. Changing the name, rather infelicitously, to l'enseignement secondaire spécial, Duruy otherwise retained the substance of Rouland's program, with its emphasis on the applied sciences, mathematics, economics, bookkeeping, history, and geography, and various manual and laboratory exercises, including drafting, to "train the mind and the eye," but no shop work. He also established a training school for professors at Cluny (Saône-et-Loire) and three special lycées in Alais (Gard), Mont-de-Marsan (Landes), and Pontivy (Morbihan).[4]

Although the *Corps législatif* voted unanimously for enseignement spécial and Cluny, it allocated no funds, and Duruy had to use all his ingenuity to make the program work. He was faced with the indifference of many industrialists and the hostility of traditionalists. The latter resented the intrusion of vulgar modern studies into the lycées and the bestowal of the prestigious *agrégation* on the "cow-fanciers" at Cluny, an insult, they thought, to the Ecole normale supérieure de Paris, the professors' training school for the lycées and universities. Supporters of modern studies, on the other hand, were disappointed that the four-year special program led only to a terminal *diplôme*, while the seven-year classical program led to the high-status *baccalauréat*, which opened the way to higher education. Because of the great prestige of classical studies, many parents of modest means preferred that their sons spend two or three years studying Latin, however useless, rather than modern studies.[5]

Duruy was aware of these problems but predicted that the progress of industry and science would gradually make the program more attractive to the business classes. He hoped "to build a bridge over the abyss ... between classical studies, with 44,000 students, and the primary system, with five million," which would open the way someday to a new cultural synthesis suitable to modern industrial democracy.[6] As long as Duruy remained in office, l'enseignement spécial prospered, but the Moral Order of the 1870s and the Third Republic that came into being later in the decade were less sympathetic to his idea. Critics of the program concentrated on attacking Cluny; in 1872 it lost its special connection with the ministry in Paris and was placed under the rector of the Academy of Lyon, who neglected it. Governments failed to assign professors to the school on time, programs and equipment became outdated, graduates had difficulty obtaining good positions, and the school experienced severe crises of morale.[7]

During the early 1880s Jules Ferry, the education minister, decided to place the special program on a level of equality with classical studies. This meant doing away with the old diploma and introducing the *baccalauréat de l'enseignement spécial* (later *moderne*). Accordingly, the decree laws of 1886 and 1891 lengthened programs from four to seven years and eliminated practical exercises in favor of a general culture based on the French rather than the Latin classics. In theory, then, *moderne* was the equal of classical studies, but this was far from the case in practice, and *moderne* continued as a kind of orphan child in secondary education until it disappeared in the reform of 1902. Moreover, in 1891 Léon Bourgeois, minister of public instruction, closed Cluny without further ado, transferring its students to the science faculty at Lyon. Cluny was converted into a school for foremen, and a few years later into an Ecole d'arts et métiers, which it remains to this day.[8]

The Social Origins and Careers of Students in l'Enseignement Spécial

Despite many obstacles l'enseignement spécial did manage to grow over the years. In 1865 it had 16,882 students; in 1876, 22,708; and by the 1880s, around 30,000, or almost a third of all secondary students in France and almost half of those in the municipal colleges. Most of the growth in secondary school attendance during these years resulted from the special program. The main problem remained the very high dropout rate. As almost one fourth of the students left each year during the four-year program, many municipal colleges did not even have a fourth year of studies (in which science was most emphasized), and most of the ones that did had only a few students. In 1879, of 23,000 students in enseignement spécial, 652 attended fourth-year programs, and only 563 passed the examinations for the special diploma.[9]

The clientele of enseignement spécial comprised mainly the sons of small manufacturers and businessmen (15 percent); of foremen, skilled workers, and artisans (24 percent); of farmers (23 percent); and, for the rest, mainly of shopkeepers, clerks, and teachers.[10] In terms of careers, students in the special program tended to follow in the footsteps of their fathers, but there were some exceptions: only one third of the students became manufacturers, skilled workers, foremen, and managers in the years after leaving school, whereas almost two fifths had come from families in these occupations. While farmers' sons accounted for almost one fourth of the students, less than one fifth returned to agriculture. Only 14 percent were the sons of clerks, office workers, and teachers, yet 30 percent found jobs in these fields after leaving school.

L'enseignement spécial tended therefore to drain young men from economically productive to semi- and nonproductive fields, though to a lesser extent than classical studies did. One reason was that most of the students remained in the program for only a year or two, receiving essentially an advanced primary education that qualified them only for positions as petty employees. Second, students in enseignement spécial soon realized that the real purpose of the lycée was the preparation of future elites for the government schools and for the liberal and administrative professions; hence lycée officials and professors easily persuaded the better pupils to transfer to classical studies.

Although the special program was not a complete failure, it was weakened from the start by its general nature, its inferior diplomas, the indifference of industrial groups, and the hostility of traditionalists. So it is difficult to criticize the Republic for its efforts to upgrade *spécial* into *moderne* in order to give it a chance to compete with classical studies. Duruy's old special program was retained in the revived system of higher primary schools and in the new lycées for young women. In the reform of 1902 *moderne* was replaced by a modern language and science program among the four *baccalauréat* options (option D).

The Beginnings of Technical Education, 1850–1880

The industrial boom of the 1850s saw the rapid growth of the railway, machine-construction, and metallurgical industries. The production process in these industries was based on an increasingly complex technology, which stimulated demand for engineers, draftsmen, and workers who could read instructions and blueprints. Industrialists, however, were uncertain as to how to procure the skilled personnel they required. Some hoped for a reform of the conditions of apprenticeship, while others attempted to recruit skilled men from the smaller companies and shops, which provided more varied training on the job than did the highly specialized larger companies. A few opened factory schools (Schneider-Creusot, Fives-Lille), while others looked to the state to cooperate with local officials and manufacturers in support of municipal vocational schools. Arthur Morin and Henri Tresca, directors of the Conservatoire des arts et métiers of Paris, advocated expansion of the technical and vocational schools under the Ministry of Industry and Commerce, but in the early 1860s that ministry was allocated only 150,000 francs annually for its schools. In any case most industrialists preferred a laissez-faire approach to the instruction of young workers and foremen.

The skepticism in industrial circles toward professional schools, prevalent in France during the middle decades of the century, was reflected in the writings of leading social thinkers. For example, Anthime Corbon, a former worker and later a republican senator, wrote a book entitled *L'Enseignement Professionnel* (1859) in which he argued that true professional education could not be separated from the workshop. Schools were necessary only to provide an elementary education and some exercises useful in training "the hand and the eye," as preparation for apprenticeship in a trade. Frédéric Le Play, a polytechnicien, distinguished between the culture of the *Université*, which prepared the bourgeoisie for the upper civil service and liberal professions, and that of the workshop, which prepared men for productive work. He and Corbon agreed on the absurdity of removing a young man from home and shop in order to place him in the artificial environment of the school, where he was taught only abstractions and acquired the mentality of a bureaucrat. They believed that the school-trained industrial manager was just another functionary, having no real understanding of the realities of production or the psychology of the worker. Such managers were the tools of the bosses, creating anarchy all around them.[11]

The Investigation into Technical Education, 1863–1864

In 1862 Arthur Morin and Henri Tresca wrote a report on the London exposition followed by a larger work, *De l'organisation de l'enseignement industriel et de l'enseignement professionnel en France*. They advocated establishing a three-tiered system of public technical schools under the Ministry of Industry and Commerce, which they called the *Université du Travail*, an idea they had apparently borrowed from the Société des ingénieurs civils, founded in 1848. This included apprenticeship schools for workers, middle-level technical institutions like the Arts et Métiers, and higher schools of engineering such as the Ecole centrale for the sons of industrialists and businessmen.

Morin and Tresca believed that only the Ministry of Industry and Commerce could provide the necessary state support while leaving businessmen free to oversee the professional schools at the local level. They insisted that industrial education should include manual work and rejected the idea that *déclassement* was an evil, arguing that the worker should benefit from industrial education in order to advance himself professionally. Citing the role of the Ecoles d'arts et métiers in transforming workers into technicians and engineers, they argued that because the middle classes avoided industry in favor of the liberal professions, industry had no choice but to recruit its managers and technicians from the working classes.[12]

In 1863 Eugène Rouher, minister of Industry and Commerce, organized a major conference on professional education. The commission included Morin, Tresca, Chevalier, Le Play, Mérimée, and Perdonnet (director of the Ecole centrale), and the industrialists

Jean Dollfus of Mulhouse, A. Dufour of Lyon, and Schneider of Le Creusot. The educational experts were, as usual, divided over the most suitable clientele for industrial education and the necessary amount of manual training. Some supported Morin and Tresca's idea of a *Université du Travail*, but most favored individual initiative, the role of the state being reduced to encouragement. All agreed, however, that insofar as the state did get involved in technical education, the Ministry of Industry and Commerce, not the Ministry of Public Instruction, was alone to be trusted.[13]

The commission reported in 1865, with Morin the *rapporteur*. Thirty witnesses had testified, of whom only five were industrialists. The commission recommended the establishment of a Conseil supérieur de l'enseignement technique to advise the minister of Industry and Commerce and called for an informal system of inspection of technical schools and the certification of teachers. The improvement of apprenticeship was to be left to local initiative, although the state was asked to tighten up the law of 1841 on child labor. Though most witnesses had stressed the weaknesses in working-class instruction, the commission confined itself to recommending an increase in the number of Ecoles d'arts et métiers and of higher business and technical schools for industrialists, whom they felt were all too often ignorant of industrial techniques.[14]

The commission's recommendations were not acted upon. The Empire shifted its attention to Duruy's enseignement spécial for the rest of the decade. In the 1860s and 1870s a few écoles professionnelles were established locally. Paris established several municipal professional schools modeled on the Ecole Turgot: Colbert, Diderot (oriented toward industry), Dorian, Lavoisier, and J. B. Say. In 1874 the Commerce ministry finally established the Conseil supérieur de l'enseignement technique. But technical and professional education remained essentially as Morin and Tresca had described it in 1863: "For an 'industrial army' of 1.2 million there is only one corporal for eighty privates and very few captains."[15] The idea of the *Université du Travail* lay dormant for 15 years, only to rise again in the struggle between the Ministries of Industry and Commerce and of Public Instruction during the last two decades of the century.

The Third Republic and the Rise of Technical Education, 1879–1914

Though it is generally accepted that the Third Republic neglected technical education, the regime actually experimented extensively in the field during the generation before 1914. The Ministry of Public Instruction re-created the higher primary schools, maintained a modern secondary alternative to classical studies in the lycées and collèges, and encouraged the development of applied science and engineering institutes in the universities. The Ministry of Industry and Commerce directed a small but growing system of technical and vocational schools designed to prepare young people for business and industry. However, the development of parallel systems of professional education under separate ministries did not always take place smoothly; each ministry had its own definition of what was then called "la formation professionnelle," and each sought exclusive control over intermediate technical instruction in France. In this section we shall deal with the quarrel between the two ministries and with the expansion of technical and professional schools under their control.

In the process of introducing compulsory, free, and secular primary education during the early 1880s, the Third Republic revived the higher primary schools. Until the 1930s these were the high schools of the common people and, as such, constituted the starting point for all projects to democratize secondary education in France. The higher primary schools instructed the sons of artisans, workers, and farmers, aged 12 to 15, in the advanced three R's, elements of applied science and mathematics, French history, geography, and civics, plus several hours per week of manual work in shop courses.[16]

The Ministry of Public Instruction also established a number of specialized higher primary apprenticeship schools for the sons of industrial workers and foremen. These Ecoles manuelles d'apprentissage, later renamed Ecoles pratiques de commerce et d'industrie, taught notions of the applied sciences and mathematics and the liberal arts plus a great deal of vocational and shop instruction. The four Ecoles nationales professionnelles at Armentières, Vierzon, Voiron, and, somewhat later, at Nantes were regional boarding schools training foremen and shop supervisors.[17]

The Ministry of Industry and Commerce quickly claimed jurisdiction over the Ecoles pratiques and the Ecoles professionnelles as schools training for business and industry. This led to several years of squabbling between the two ministries and finally to a government-imposed compromise called the *Condominium,* in which the two ministries shared jurisdiction over the Ecoles pratiques and Ecoles professionnelles. Under this arrangement Public Instruction had the financial authority, while Commerce had the right of inspection and a veto over programs and personnel. The result was delay and sometimes deadlock on the question of programs and especially on the amount of vocational and manual training to be given in the schools.

In the public debate that accompanied the quarrel, leading figures in the Ministry of Public Instruction, notably Jules Ferry, minister from 1879 to 1885, and Ferdinand Buisson, director of the division of primary education in the ministry from 1879 to 1896, emphasized that the French productive sector was very broad, including thousands of small firms, shops, and farms as well as big business and industry. They believed that the new system of higher primary schools should avoid a strictly vocational approach, concentrating instead on a general modern culture designed to educate workers and farmers "capable of learning on the job, of adapting to technological change, of advancing themselves according to their lights, and, as men and citizens, of understanding their rights and duties in a democratic society."[18]

Spokesmen for the Commerce group, the deputies Placide Astier and Modeste Leroy, industrialists such as Jules Siegfried, and Gustave Ollendorf, director of the technical education division in the Ministry of Industry and Commerce, spoke often of a "formation professionnelle," a combination of practical and classroom training that would provide the future worker with the skills needed to advance himself on the job and furnish industry with the technical manpower necessary in an age of struggle. They attacked the higher primary schools under the Ministry of Public Instruction as "second-class lycées," draining the better students from the productive sector "to form an army of déclassé intellectuals and clerks."[19]

The Commerce group wanted the Condominium ended and the Ecoles pratiques and Ecoles professionnelles transferred to the Commerce Ministry. They also proposed the conversion of all higher primary schools located in industrial areas into Ecoles pratiques and the transfer of professional schools scattered under other ministries (Agriculture, Fine Arts, Public Works, and so on) to Commerce, which would then head a kind of *Université du Travail* alongside, but separate from, the Napoleonic *Université.* To win support for these projects, they played heavily on patriotic themes, arguing that the Condominium was delaying progress in technical instruction that was vital to meet the demand of heavy industry for skilled men brought about by advances in industrial technology and German competition.[20]

In this way there emerged a well-organized technical education lobby centered in the Commerce Ministry's Conseil supérieur de l'enseignement technique. This council included representatives from business, industry, and engineering associations who worked closely with the permanent officials of the technical education division of the ministry and with a series of able ministers of commerce, notably Alexandre Millerand (1899–1902). Almost one fifth of the deputies in Parliament in 1912 (a total of 114 deputies, mostly of the right-center) listed technical education as one of their highest priorities.[21] Liaison among this large but amorphous parliamentary group, the industrial associations, and the ministry was realized through the Conseil supérieur de l'enseignement technique and later also through the Association française pour le développement de l'enseignement technique, founded in 1902 by Modeste Leroy and other politicians and industrialists to direct the campaign for the advancement of technical education in France.[22]

In 1892 the Commerce group had a major success when it persuaded Parliament to place the 20 existing Ecoles pratiques and all other higher primary schools in which the teaching was "principally industrial or commercial" under the Ministry of Industry and Commerce. The four Ecoles professionnelles remained under the Condominium until 1900, when the Chamber of Deputies voted to transfer them, along with six vocational schools, to Commerce. These reforms accompanied a rise in the education budget of the Ministry of Industry and Commerce from 2 to 10 million francs annually between the early 1890s and 1914.[23]

The end of the Condominium, brought about by the parliamentary decisions of 1892 and 1900, gave Commerce a small but complete system of schools ranging from the Ecoles pratiques and Ecoles professionnelles to the intermediate Ecoles d'arts et métiers to such higher engineering schools as the Ecole centrale. But it left the status of the higher primary schools unclear. The law of 1892 allowed the conversion of any such school into an Ecole pratique when its programs were "principally industrial or commercial," but this depended upon the assent of the municipality in question. During the following years, the Commerce Ministry successfully persuaded several industrial cities to convert their higher primary schools into Ecoles pratiques, which exacerbated tension between the two ministries and placed Public

Instruction in a very difficult position. If it introduced practical and professional programs into the higher primary schools, it risked losing them to Commerce; if it did not do so, it was attacked by the Commerce group for the irrelevance of its programs to the needs of the modern world and the arms race with Germany. The Chamber of Deputies partially solved this problem in 1906 and 1907 when it called upon the Ministry of Public Instruction to orient the higher primary schools in a more professional direction and raised the annual budget for manual work in these schools from 6,000 to 300,000 francs.[24] This, and the steady rise in the Commerce budget, allowed each system to grow independently of the other and ended the worst aspects of the quarrel between the two ministries.

The Ecoles Pratiques de Commerce et d'Industrie

The Ecoles pratiques dated to 1880, when a budget was established to create apprenticeship schools, the Ecoles manuelles d'apprentissage, to train skilled workers and office employees for business and industry. They developed slowly, hindered by the quarrel between the two ministries and by the Condominium of 1886. When they were placed under Commerce in 1892, there were about 20 of them, with 1,300 students. Renamed Ecoles pratiques de commerce et d'industrie, they grew rapidly to 70 in 1913, with 14,766 students, of which one fourth were for girls (table 3.1). In addition, there were 20 closely related vocational schools in Paris, with 3,500 students. All were day-schools designed to provide technical and vocational instruction for students aged 12 to 15. The students spent an average of 25 to 30 hours per week in the classroom learning the advanced three R's, moral and civic instruction, applied geometry (especially mechanical drawing and industrial design), some algebra, and the elements of the industrial sciences (mechanics, electricity, chemistry). Twenty-five to 30 hours were also spent in the workshops, the nature of which varied according to local needs. The Ecole Vaucanson of Grenoble, for example, concentrated on electricity and glovemaking; Montbéliard on automobile production and precision instruments; Bordeaux on agricultural chemistry; and Brest on marine mechanics.[25]

As for recruitment, the director of the Ecole pratique of Saint-Etienne reported that after the foundation of the school in 1880, most of the students had been either the sons of workers or the rejects of other institutions. By 1900 "foremen, shop supervisors, managers and owners, and even a few high-ranking industrialists have begun to

Table 3.1
French intermediate, technical, and higher primary schools and students, 1895–1913

	1895		1905		1913	
	Schools	Students	Schools	Students	Schools	Students
Ecoles pratiques	25	2,607	50	10,342	70	14,766
Ecoles nationales professionnelles	3	750	4	1,327	4	1,800
Ecoles d'arts et métiers	3	900	5	1,500	6	1,800
Ecoles primaires supérieures	260	33,000	332	45,000	450	55,000
Totals	291	37,257	391	58,169	530	73,366

SOURCES: P. Astier and I. Cuminal, *L'enseignement technique, industriel et commercial en France et à l'étranger*, 2nd ed. (Paris, 1912), pp. 218–222; Ministère du Commerce, de l'Industrie, des Postes et des Télégraphes, Conseil supérieur du travail, *Enquête récente sur l'enseignement professionnel en France, Rapport de M. Briat au nom de la commission permanente: procès-verbaux des seánces* (Paris, 1905), pp. 51–78.

send their sons." The director at Rouen reported that in 20 years his school had grown from 48 to 170 pupils and that industrialists and managers had begun to show interest: "The appearance of these students at the school is noteworthy: it shows that some non-working-class parents are not letting themselves be influenced by the all too widespread prejudice against manual work in the education of their children." [26]

An analysis of a sample of 3,670 students of the Ecoles pratiques during the 1890s revealed that 46 percent (1,689) were the sons of workers and only 3 percent (112) of manufacturers and managers; one fifth were the sons of artisans, shopkeepers, and farmers (766); and slightly less than one fourth were the sons of clerks and petty officials in business and government (844, or 23 percent). Most of the students came from an urban-industrial milieu and returned to it after completing their studies (table 3.2). [27]

The placement of graduates improved steadily during the 1880s and 1890s. In the early 1880s employers, foremen, and workers were often suspicious of school-trained personnel and were hesitant to hire them. But by 1900 the directors of Lille, Rouen, Saint-Chamond, and Saint-Etienne were reporting that placement was easy and that their graduates were earning good salaries. The director at Saint-Etienne said in 1900 that at the age of 20 the average graduate was earning five francs per day, "vastly superior to the one franc per day earned by young men their age who learn their craft on the job." He reported that one fifth of the 730 graduates over the past 20 years had already become independent owners, entrepreneurs, or supervisors and fore men in industry.[28]

A study of the early careers of 4,603 graduates during the 1890s shows that more than half (52 percent or 2,373) became workers, 20 percent (941) clerical employees in business and industry, and 5 percent (230) employees in government services. Eight percent obtained further education either in an Ecole d'arts et métiers (196) or in other schools (181) (table 3.3).[29] Almost three fourths of the graduates went into business and industry immediately on finishing their studies, and most of those who continued their studies returned to industry later on. Generally, the students were encouraged to find jobs as soon as they had completed their studies.

Industrialists liked the Ecoles pratiques because they were primarily local institutions clearly oriented toward business and industry. Such men dominated the boards of trustees attached to each school and had a considerable say in formulating their programs. Emile Montgolfier, director of the metallurgical and arms company Aciéries de la Marine et d'Homécourt, told the Conseil supérieur de l'enseignement technique in 1904 that "the Ecoles pratiques generally train our best workers and foremen; we hire many of their graduates, and, if they are intelligent, they move up quickly in our hierarchy." Armand Peugeot, director of the automobile company bearing his name, said that he employed 50 graduates of the Ecole pratique de Montbéliard in his factory at Audincourt and spoke of the important services that the school had rendered to the automobile industry in the area.[30]

The Ecoles Nationales Professionnelles

The three schools at Armentières, Vierzon, and Voiron were originally called Ecoles primaires supérieures et professionnelles. They were established by the Ministry of Public Instruction in the early 1880s as part of the experimental "groupes scolaires" running from preschool through the elementary grades and culminating in the higher primary professional sections for young men aged 12 to 15, in "preparation for life and work in a democratic society." As few local students actually moved from the primary to higher primary professional sections, the Ecoles professionnelles gradually became boarding schools. They recruited from the surrounding regions, and most of their students held state scholarships while training for careers as foremen and shop supervisors in the mechanical branches of industry. Their programs included chemistry, physics, mechanics, mathematics, mechanical drawing, and shop work and ranged in difficulty between those of the Ecoles pratiques and the Ecoles d'arts et métiers. Their brighter students prepared for the latter schools, the remainder going directly into industry as technicians.[31]

In 1886 the Ecoles nationales professionnelles were placed under the Condominium and soon became the focus of an acrimonious dispute between the two ministries. Public Instruction felt that its experiment in education from preschool to professional school had not been allowed sufficient time and had been undermined by Commerce. The latter in turn used the failure of the experiment to demonstrate the incapacity of Public Instruction in the field of professional training and the need to convert such schools into strictly technical institutions under its own direction. When in 1898 the two ministries prepared to open a fourth school at Nantes, there were so many delays that in 1900 Parliament decided to break the knot and place the schools exclusively under Commerce.[32]

In a sample of 1,267 students at Voiron in the 1890s, one half were the sons of either manufacturers and managers (15 percent), artisans and shopkeepers (23 percent), or workers and foremen (12 percent). Seventeen percent were the sons of farmers (table 3.2). In terms of career choices, of 774 graduates of Armentières, Vierzon, and Voiron from 1896 to 1899, almost half started as skilled workers and foremen, one tenth as employees and clerks in business and industry, and an equal percentage as small owners, artisans, and shopkeepers. Only 2 percent returned to the farm. One fourth continued their studies, usually in an Ecole d'arts et métiers, and normally they later returned to industry in supervisory roles (table 3.3).[33] While half of the students of the Ecoles professionnelles came from industrial or commercial backgrounds, frequently from small firms and shops, three fourths went to work in larger companies where, as skilled workers, they could become foremen, supervisors, technicians, and engineers through promotion on the job.

Graduates of the Ecoles professionnelles were much in demand in industry. In 1899 the director of the school at Vierzon reported: "I receive many letters from industrialists and almost all are very satisfied with our young men. Hardly a fortnight goes by that I am not contacted about job openings, and it is very rare that a manufacturer, having employed one of ours, does not hire others."[34]

Table 3.2
Professions of fathers in selected schools under the Ministry of Commerce

Profession	Ecoles pratiques de commerce et d'industrie, 1890s		Ecoles nationales professionnelles, Voiron 1886–1899		Ecoles d'arts et métiers, 1800–1890	
	(n = 3670)	percent	(n = 1267)	percent	(n = 503)	percent
Business and Industry						
Manufacturers and managers	112	3	200	15	98	19
Clerical employees	534	15	218	17	11	2
Workers and foremen	1,689	46	150	12	73	15
Artisans and shopkeepers	683	19	288	23	115	23
Government Services	310	8	113	9	80	16
Farmers	83	2	213	17	25	5
Miscellaneous	71	2	74	6		
Unknown	188	5	11	1	101	20

SOURCES: Ministère du Commerce, de l'Industrie, des Postes et des Télégraphes, Direction de l'Enseignement technique, *L'Enseignement technique en France, Etude publiée à l'occasion de l'Exposition de 1900*, 5 vols. (Paris, 1900), vols. 2–3: *Ecoles pratiques de commerce et d'industrie*; *AN*, F 17 14354, Voiron, 1900, and F 17 14350, Report of M. Leblanc, 1900; C. R. Day, "The Making of Mechanical Engineers in France: The Ecoles d'Arts et Métiers, 1803–1914," *French Historical Studies* 10 (1978), 439–60, and "Education for the Industrial World," in R. Fox and G. Weisz, eds., *The Organization of Science in France* (Cambridge: Cambridge University Press, 1980), pp. 127–153.

Table 3.3
Early career choices of graduates of post-primary modern and technical schools

Profession	Ecoles primaires supérieures, 1902		Ecoles pratiques, 1890s		Ecoles nationales professionnelles, 1896–1899	
	(n = 5033)	percent	(n = 4603)	percent	(n = 774)	percent
Business and Industry						
Manufacturers, artisans, and shopkeepers	988	20	50	1	77	10
Clerical employees	1,576	31	941	20	73	10
Workers and foremen	1,010	20	2,373	52	365	47
Government Services	282	6	230	5	30	4
Farmers	678	13	165	4	17	2
Ecoles d'Arts et Métiers	256	5	196	4	160	21
Other Schools	243	5	181	4	42	5
Miscellaneous	—	—	365	8	—	—
Unknown	—	—	102	2	10	1

SOURCES: See the preceding table. On the Ecoles primaires supérieures, see *AN* F 17 14350, Ministry of Public Instruction (M. Gasquet, minister), 1902.

The Ecoles Primaires Supérieures

It was an article of faith in the Commerce Ministry that the higher primary schools and the modern program in the secondary schools were diverting energy and talent away from industry and that the higher primary schools located in industrial centers should be converted into Ecoles pratiques. Defenders of the higher primary schools replied that those located in industrial cities did provide shop instruction (up to 15 hours per week) but that there was little demand for such facilities in the smaller cities and towns where most of the schools were located. In these localities small firms and farms abounded, and young people could learn their skills on the job. They cited official statistics indicating that slightly more than 300,000 workers (of a total of 3 million in the labor force) worked in 151 companies having more than 1,000 employees, while 807,000 worked in 489,000 firms having 1 to 4 employees.[35] In such a diversified economy the school could not provide specialized vocational instruction for each student; the goal was to provide a good general modern instruction that would encourage a more intelligent approach to the direction of various enterprises.

In 1900 the Ministry of Public Instruction published statistics on 61,686 former students of the higher primary schools during the 1890s: these indicated that almost one third (30 percent) of the fathers worked in industry, mainly as workers, nearly one fourth (23 percent) in business, mainly as clerks and shopkeepers, and the same percentage (23 percent) in government, as employees and teachers. Seventeen percent were farmers. A 1902 Ministry of Public Instruction study on the early jobs obtained by 5,033 graduates revealed that one fifth became small manufacturers, artisans, and shopkeepers, another fifth workers and foremen, and almost one third clerks and employees in business and industry. Only 6 percent became government clerks and teachers, and another 10 percent continued their education, half of them in the Ecoles d'arts et métiers. Hence the number employed in business and industry (including small shops) rose from slightly over half of the fathers to about three fourths of the sons (counting those who later graduated from technical and professional schools). The number of government clerks and teachers decreased from 23 percent (fathers) to 6 percent (sons), while agriculture fell from 17 to 13 percent (see table 3.3).[36] Even though many of the enterprises concerned were small in scale and many graduates did obtain jobs as clerks, usually in industry rather than government, there is no evidence that the higher primary schools drained from the productive sector into government bureaucracies. The fact that some continued their studies in the Ecoles d'arts et métiers is important, for the entrance examinations for these schools were difficult, including written, oral, and manual exercises, and poorly prepared students could not have passed them.[37]

The higher primary schools provided for a broad range of educational opportunities and complemented the Ecoles pratiques and Ecoles professionnelles. They drew many students from small family firms and farms, and about one third of the graduates returned to them. Although they were criticized for this, the small firm was not necessarily conservative or technically backward. Moreover, the comparable figures were 16 percent for the Ecoles pratiques and 21 percent for the Ecoles professionnelles. The higher primary schools were also criticized for not possessing shop facilities, but in areas with no clear industrial concentration, such facilities would have served little purpose. The higher primary schools were not second-class lycées draining able students from productive to nonproductive enterprise (although this was true of *moderne*) but rather were adequately suited to the needs of the many businesses in France that had no need for highly trained specialists.

Technical Instruction on the Eve of World War I

The impression among historians of the inferiority of French technical education has been based mainly on a series of studies and reports sponsored by the Ministry of Commerce from the late 1890s to 1905 and on a book published in 1908 (revised in 1912) by Astier and Cuminal on technical education in France and abroad.[38] J. P. Guinot drew heavily upon the reports of the Ministry of Industry and Commerce and of Astier and Cuminal in his book on working-class education published in 1946.[39] Such works provided abundant information on the state of technical education in France, but they may have exaggerated somewhat the inferiority of French to German technical education, perhaps to win greater government and public support for the schools under Commerce.

In their 1912 edition, for example, Astier and Cuminal stated that there were only 25,000 students in technical schools in France. They contrasted this with 500,000 young people in Germany receiving technical and commercial training, 46,000 in Belgium, and 22,000 in Denmark. However, their figure of 25,000 (20,000 in the Commerce system and 5,000 in mainly private technical schools) was outdated by 1912. By then the rapidly growing Commerce system alone had more than 25,000 students and grew to more than 28,000 in 1914. In addition, there were 5,000 students in private technical schools, 5,000 in schools under other ministries (overlooked by Astier and Cuminal), and more than 50,000 in the higher primary schools, of whom 7,400 were enrolled in technical programs, while some of the remainder took up to 15 hours of shop courses and practical work.[40] Finally, there were 70,000 students attending continuing education courses of a technical nature. On the eve of the war, therefore, at least 150,000 students were enrolled in higher primary and intermediate technical schools and programs in France, a considerably higher figure than that given by Astier and Cuminal.[41]

Astier and Cuminal's statistics for Germany were heavily based on workers enrolled in continuing education courses. Apart from this, their figures show no great disproportion between the two countries, particularly when we remember that Germany had almost twice as many young people under 21 as did France. In 1899 the *Bulletin de l'enseignement technique* acknowledged that "German superiority does not lie at the level of higher technical education or in business education, but at the level of working-class instruction."[42] In 1909 Camille Cavallier, director of the big metallurgical works at Pont-à-Mousson in Lorraine, told a congress of industrialists that the main problems facing French industry were expensive resources and labor. He said that "there was no scarcity of engineers in France," but he did think that the number of trained workers and foremen could be enlarged.[43] Although there is little doubt that Germany was "ahead" in the apprenticeship training and in the continuing education of adolescent workers, it is not clear whether such instruction was necessary. Given the simplicity and repetition of even skilled tasks in the modern factory, it is possible that the hidden purpose of such education was to turn out docile workers and stream them early so as to avoid *déclassement*. But as Morin, Tresca, and others pointed out, a certain amount of *déclassement* was necessary, even desirable, to meet the objectives of a democratic society and replenish technical elites.

It appears, then, that French technical education around the turn of the century was meeting the needs of French business and industry for skilled personnel. In 1906 the president of the Société des ingénieurs civils de France said that the balance of supply and demand for technical personnel in France was satisfactory.[44] In 1913 Modeste Leroy, long the chief Cassandra of the Commerce group, said:[45]

I am proud to state that technical education, which was almost unknown in France twenty years ago, has conquered today a splendid place in the sun: governments and parliaments neglect no occasion to show their good will, and the municipalities and departments compete with each other to work with the technical education department of the Ministry of Commerce [in establishing technical schools].

From the 1880s to 1914 the number of intermediate technical schools in France steadily increased, and the schools were modernized and reequipped. Several training schools for teachers in technical instruction were combined into a single Ecole normale d'enseignement technique near Paris in 1912.[46] Most of the schools under Commerce were closely linked to industry through the various councils of the ministry, the boards of trustees of the schools, and the alumni and professional associations. Thus by the eve of World War I, the technical schools under Commerce and other ministries appear to have adequately met the demand of business and industry for skilled men, leaving the Ministry of Public Instruction to provide an intermediate modern education in the higher primary schools for young people who needed neither vocational nor classical instruction. As we shall see in Chapter 11, the excellent performance of French industry in adapting to a war of attrition during World War I is evidence of the accuracy of this hypothesis.

4

Intermediate Technical Education in France Since World War I

Now there was a poor wise man, and he by his wisdom delivered the city; yet no man remembered the same poor man.

Then said I, Wisdom is better than strength: nevertheless the poor man's wisdom is despised, and his words not heard.

Ecclesiastes 9:15–16

The interwar period saw few major changes in intermediate education in France. Several reforms, however, are worth mentioning, notably the Astier Law on vocational continuing education of 1919, the transfer of the Commerce system of schools to the Ministry of Public Instruction in 1920, the abolition of tuition in the secondary lycées and collèges in the late 1920s, and the beginnings of the assimilation of the higher primary schools to them.

The Astier Law was the culmination of nearly 20 years of campaigning by the Commerce lobby. The preamble stated: "Because of the changing needs of industry and the role of business and industry as a source of national wealth, the state must take the place of private associations in providing professional education for the sons of workers."[1] The law made professional training compulsory for all young workers until the age of 18, four years beyond the legal age, requiring them to attend vocational classes for at least 4 hours per week and for a minimum of 100 hours per year. Such classes were to be established in factories, businesses, or schools, and attendance over three years led to a certificate of professional aptitude. Local patronage boards were drawn from labor and employers' groups.

The Astier Law applied only to the larger companies, and even then governments left enforcement of the law to private associations; they failed to provide adequate funds for the creation of new courses or to oblige reluctant employers and public authorities to establish them. Yet the law did pave the way for other reforms, which laid the basis for future expansion. In 1925 the government created agencies to supervise apprenticeship, established orientation and placement services, and imposed an apprenticeship tax on wages, the revenue to be used to train young workers. Many factories set up their own training programs in order to be excused from the tax. A teacher-training school, L'école normale nationale d'apprentissage (ENNA), was established in 1934 to prepare instructors in vocational education. In 1938 the minimum requirement for the professional education of young workers was raised from 100 to 150 hours per year, and employers were obliged to present their apprentices for a public examination. That year also saw the establishment of centers of professional schooling (centres d'apprentissage), which later evolved into the collèges techniques.[2] By 1940 there were 40,000 candidates for the different types of certificates of professional aptitude. A decade earlier this would have been a sign of progress, but on the eve of the war it was too little, too late. In the midst of large-scale unemployment during the 1930s, France had to import skilled workers from abroad.

After World War I a group of educational reformers called the *Compagnons de l'éducation nouvelle* attacked the elitism and fragmentation of French education and called for the creation of a single, comprehensive secondary school, the *école unique*. This meant raising the age for leaving school to 16, abolishing school fees in secondary education, and uniting the various lycées, collèges, and higher primary and technical schools scattered under several ministries into a comprehensive secondary school under the Ministry of National Education. The adherents of reform were divided on the form that the new school should take: some wanted an American-style high school, with all students under one roof until the age of 16; others thought in terms of a common junior high school (ages 12 to 14), with tracking into classical, modern, and technical streams in senior high school (15 to 17); while still others wanted only a common program (*tronc commun*) for the early years of high school, to be taught separately in the various types of secondary schools.[3]

These divisions weakened the reform movement. Under the Popular Front government, Jean Zay, the education minister, presented a bill to Parliament in 1937 that attempted to establish a comprehensive school in the early high school years and unify French secondary education under his ministry, but the combined opposition of educational pressure groups prevented its passage.[4]

The only progress toward the unification of French schools,

scattered under the jurisdiction of several ministries, came in 1920 when the government of Alexandre Millerand transferred the technical schools located under the Ministry of Industry and Commerce to the Ministry of National Education. The Commerce lobby, represented by the Association pour le développement de l'enseignement technique, vigorously protested that the schools would lose their identity in the huge bureaucracy of the public education system.[5] In response Millerand created a special technical education division within the ministry with its own under secretary of state and its permanent director, who was equal in rank with the directors of the primary, secondary, and higher-education divisions of the ministry.

By the beginning of the 1920s, therefore, the Ministry of National Education had three kinds of intermediate schools under its jurisdiction, each with its own programs, diplomas, faculty, and clientele: the lycées and collèges, with their classical and modern *baccalauréat* options; the higher primary schools; and the intermediate technical schools, with their *brevets* and *certificats*. These three types of schools remained generally separate from one another, though some bright students of modest origin did move from *primaire* to *secondaire* and thence to the universities and sometimes even to the *grandes écoles* via state scholarships. Enrollments in the classical secondary schools remained stagnant from 1880 to World War II. Growth was restricted to female secondary education and to the higher primary and intermediate technical schools.[6]

The popularity of the higher primary schools derived from their close ties with the primary system; their programs were oriented toward the needs of the community, and they were more accessible in country areas than were the lycées and collèges. Taught by graduates of France's normal schools possessing the advanced teaching certificate (the *brevet supérieur*), these schools were considerably better than their critics in classical and technical education acknowledged. But their flourishing enrollments made them vulnerable to the ambitions of their rivals. During the interwar years the Ecoles pratiques absorbed 35 higher primary schools located in industrial and commercial areas, and the municipal colleges managed to annex about 100.[7]

Located in smaller cities and towns, the municipal colleges were more dependent on local support than the state lycées and were threatened by declining enrollments. By the 1920s the modern programs of the colleges and those of the higher primary schools were virtually indistinguishable. In urban areas these schools often shared the same premises, facilities, and faculties. There were only two important differences between them: the secondary curriculum led to the *baccalauréat de l'enseignement moderne*, while the higher primary program led to a certificate of studies, and students in the secondary program paid tuition fees while those in *primaire* did not. During the enrollment crisis of the late 1920s, caused by the low birthrate of the war years, the government abolished fees in the secondary schools. This and raising the age for leaving school from 13 to 14 marked the beginning of the end of the higher primary schools in most cities. With the abolition of fees in secondary schools, students in the higher primary sections could easily transfer into classical and modern programs and prepare for the *baccalauréat*. This threatened to deprive the Ecoles d'arts et métiers, which drew half their recruits from the higher primary schools, of their best source of students, leaving only the Ecoles nationales professionnelles and, to a lesser extent, the Ecoles pratiques as sources of qualified students. As we shall see in chapter 7, the decline in qualified applicants during the 1930s was a primary reason why the Ecoles d'arts et métiers sought to reduce their enrollments.[8]

The Ecoles professionnelles and Ecoles pratiques grew moderately during the interwar years. In 1914 the Ecoles pratiques had 15,000 students; in 1938 they had almost 65,000, though some of these had come from the absorption of higher primary schools.[9] The Ecoles professionnelles numbered 4 after World War I, the "quatre vieilles" at Armentières, Vierzon, Voiron, and Nantes. Several years after the war the education ministry converted 4 of the strongest Ecoles pratiques at Epinal, Lyon, Saint-Etienne, and Tarbes into Ecoles professionnelles and also assimilated the 2 écoles d'horlogerie at Besançon and Cluses to them. Suddenly, there were 10 rather than 4 schools, and this awakened fears of being swamped by the less prestigious Ecoles pratiques.[10] The alumni association of the Ecoles professionnelles grudgingly accepted the graduates of the 6 new schools into its ranks, but it used its influence to prevent the creation of any more schools. When in the late 1920s the administration opened 3 new ones, including a school for young women at Bourges, this was the last straw. The association refused to admit their graduates into its ranks, creating a schism that was not closed until 1936. In the meantime 3 more schools had been established over the protests of the association, bringing the number to 16. When the education law of 1959 merged them into the system of lycées techniques, the Ecoles professionnelles numbered 32, with 18,500 students.[11] As lycées techniques they have gradually lost their separate identity, and their alumni association now represents technicians generally.[12]

During the interwar years, therefore, France made modest gains in technical education. The number of students in the Ecoles pratiques grew from 15,000 in 1913 to 65,000 in 1938, and the Ecoles professionnelles went from 1,800 in 1913 to 9,000 in 1938.[13] However, the number of students in the Ecoles d'arts et métiers declined from 1,800 to 1,000 in 1938. There were also some gains made in continuing education; the Astier Law of 1919 set the stage for compulsory part-time education for young workers but was never fully implemented between the wars. The creation of centres d'apprentissage in 1938 for training and retraining workers was important, but these were only fully developed under Vichy and the Fourth Republic and were later assimilated to the collèges techniques.

Technical Education in France Since World War II

After the stagnation of the interwar period and the disruptions of the war and cold war years, French education began to expand very rapidly beginning in the 1950s. In 1945 there were 737,000 pupils in the secondary schools; 30 years later there were 5 million. French secondary education continued to be divided into classical, modern, and technical streams, of which the latter grew most rapidly, becoming the largest of the three by the 1970s. As in other European countries, the opening up of secondary education was soon followed by the rapid expansion of the universities in the 1960s. In 1945 there were 80,000 students in higher education; in 1976, 1,057,000.[14]

In the immediate postwar years three types of secondary schools emerged: (1) the classical lycée, which had absorbed the more important collèges communaux; (2) the collège moderne, which had grown out of the old higher primary schools, as well as some of the less significant collèges communaux; and (3) the lycées and collèges techniques, intermediate technical schools under their own semi-autonomous division within the Ministry of National Education. Vichy had created the collège technique from the Ecoles pratiques de commerce et d'industrie and some of the more technically oriented higher primary schools. In 1946 the government established a *baccalauréat technique* for the new system of lycées techniques. The Ecoles d'arts et métiers were promoted to university status in 1947, which meant that technical education had a complete system of schools and virtual parity in programs and diplomas with *classique* and *moderne*.[15]

These important reforms in the technical stream accompanied hopes of the democratization of French education. In 1944 General Charles de Gaulle, interim president of the Republic, nominated physicist Paul Langevin and psychologist Henri Wallon to co-chair a commission charged with providing an integral plan for the reform of the French educational system. The report, presented in June 1947, recommended the extension of compulsory education from 14 to 18 years and the division of the public education system into three "degrees" or cycles, as opposed to the previous levels rigidly separated from each other. The elementary cycle was to run from the Ecole maternelle to age 10; an orientation cycle and common program (*tronc commun*) followed for young people aged 11 to 15. Students going on to senior high school were to be directed into classical, modern, and technical streams but with ample means of transferring from one to the other. The third, higher-education cycle was to be divided into a two-year propaedeutic term followed by a program leading to the *licence* diploma.[16]

The purpose of the Langevin-Wallon recommendations was to open up the secondary system and close the gap between classical studies and practical education. They placed great emphasis on the common school and the orientation cycle for young people aged 11 to 15, whose progress was to be observed by teachers and a battery of trained counselors and educational psychologists. The latter, in consultation with parents, were to guide young people in the direction most suitable to their abilities. Students in the arts and sciences were to receive some practical training, while those in technical programs were to take some general courses, a flexible and open approach to learning called *polyvalence* by the French.[17]

In 1946 the Constituant Assembly grandly declared: "La Nation garantit l'égal accès de l'enfant et de l'adulte à l'instruction, à la formation professionnelle et à la culture." Yet the recommendations of the Langevin-Wallon Commission were never implemented. A law based on the commission's findings in 1948 did not even reach the Conseil supérieur de l'éducation nationale for study. In 1949 the deputy, Yvon Delbos, prepared a bill that was sent to the Conseil supérieur, which rejected it.[18] A few piecemeal reforms were subsequently introduced. The centres d'apprentissage were assimilated to the collèges techniques in 1949, and five regional Ecoles normales nationales d'apprentissage were established to train professors for the collèges techniques. The Ecole normale nationale supérieure de l'enseignement technique, established in 1912, which trained professors for the lycées techniques, was transferred to a big new center at Cachan in 1957.[19]

Otherwise the system remained as it had been before the war. Generally the children of the well-to-do attended the lycées, those of the middle and lower classes enrolled in *moderne*, and those of the working classes gravitated to *technique*. The great majority of the pupils in the latter were from families of petty employees, workers, artisans, and small farmers: 60 percent in the Ecoles nationales professionnelles, 75 percent in the collèges techniques, and 80 percent in the centres d'apprentissage. In the 1950s 90 percent of young people from the families of workers and peasants quit school at the legal age of 14 to go to work. Only 8 percent of the children of workers attended a lycée, and 3 percent went on to higher education. Conversely, 85 percent of children from the liberal professions, higher civil service, and corporation executives attended lycées, most of whom continued into higher education.[20]

With the expulsion of the Communists from the government in 1948, the Fourth Republic returned to prewar routines. Yet changes were taking place that would transform French society and French education. Postwar reconstruction and economic planning gradually redefined the structures of the French economy. The European Economic Community and the Common Market of the 1950s and 1960s forced French industry to retool and compete on the international market. The birthrate rose from a low of 531,000 in 1941 to 860,000 in 1950, which made reform necessary in order to accommodate the large numbers of young people who would begin reaching the secondary system in 1960.[21] Technical education was bound to feel the brunt of economic change and demographic growth. The division grew from 66,000 in 1939 to almost half a million in 1948 (240,000 in the lycées techniques, écoles nationales professionnelles, and collèges techniques; 200,000 more in the centres d'apprentissage and continuing education programs).[22] This sudden growth overwhelmed facilities. Old barracks were hastily converted into technical schools, and primary school teachers were given several months of crash courses in technology. Even then 50,000 young people were turned down annually for admission to the technical schools.[23]

The technical education system continued to grow rapidly throughout the 1950s, reaching 726,000 students, with 55,000 teachers, by the end of the decade, but the division continued to be underfinanced and deficient in teachers, space, and material. A series of investigations between 1955 and 1960 by the Union of Metallurgical Industries into the manpower requirements of the industry revealed that France lacked skilled personnel at all levels. The country needed 10,000 engineers per year and was producing only 4,500. The shortage of trained technicians, especially in chemistry, was even worse. The écoles nationales professionnelles, the lycées techniques, and a few other schools graduated 10,000 technicians annually, satisfying about a third of the demand. There were three engineers for every two technicians in France. Half of the technicians working in industry did not possess the appropriate educational credentials; for the most part, they had acquired their skills on the job or privately.[24] Further studies revealed that school-trained technicians were no more efficient than those who had learned on their own. This was taken to mean that technical schools should be improved and expanded rather than eliminated. Another investigation (there were many during these years) revealed that four fifths of skilled workers and two fifths of young people leaving school had obtained no vocational training at all.[25]

In 1956 Jean Berthoin, the education minister, presented a bill for the reform of secondary education to Parliament. The bill was to raise the age for leaving school to 16 and to introduce a common program (*tronc commun*) in the first two years of secondary school for young people aged 13 and 14 before streaming them into classical, modern, or technical studies. Though it fell far short of creating an *école unique*, the bill was opposed by well-entrenched interest groups, notably the secondary professors' association.[26] It did not pass until 1959 and then only with the coming of the Fifth Republic in 1958 and warnings of a potentially disastrous shortage of skilled personnel in the 1960s. The law created a "long" technical program in the lycées techniques, leading to the *baccalauréat technique* and thence into higher education, mainly to the Ecoles d'arts et métiers, and a terminal "short" program for the collèges techniques, preparing for a *certificat d'aptitude professionnelle* (*CAP*) and *brevet de l'enseignement technique* (*BET*) for those leaving school at the new legal age of 16. It also created the collège d'enseignement général from the *cours complémentaires*, a kind of postprimary extension course for rural populations. Located mainly in rural areas, these did not grow very rapidly and failed to become the comprehensive schools of the sort recommended by the Langevin-Wallon Commission 12 years before.[27]

The two-year orientation cycle created by the law of 1959 was introduced mainly into the existing lycées and collèges, where counselors tended to pass the better students directly to classical or modern studies rather than transfer them to the lycées and collèges techniques. When they did attempt to do so, the parents often refused, preferring a higher-status general education. Further, children were reluctant to leave the school and the friends they had made for an unknown and often more distant school. As Edgar Faure said, "On porte le deuil du latin en entrant dans le moderne et le deuil du moderne en entrant dans la technique."[28] The old system, in which graduates from the primary schools had been assigned directly to classical, modern, or technical schools, had worked better for *technique*.

The result was a decline in the appeal of intermediate technical instruction, especially of the lycées techniques, the principal source of technicians in France. The blow was doubly serious because the law of 1959 had merged the Ecoles nationales professionnelles into the lycées techniques.[29] To make things worse, the establishment of the Instituts universitaires de technologie (IUTs) in the mid-1960s, two-year institutes training high-level technicians and opening the way into higher education, destroyed the preparatory sections of the lycées techniques leading to the universities. By 1967 there were 15,000 empty places in the lycées techniques, and by 1970 the IUTs had only 32,000 students, where they had room for 45,000.[30] France's supply of technicians was apparently drying up just at the time of greatest demand.

The technical education division had strongly opposed the merging of the écoles nationales professionnelles into the lycées techniques. In response the Ministry of National Education did away with the technical education division in 1964, which had been in existence in one form or another since 1832, assigning the schools to the secondary and higher-education departments in the ministry. Defenders of this decision argued that it was necessary to terminate ghettos within the public education system, but this did not assuage the fears of technical education groups that the dispersal of technical schools in the giant educational bureaucracy of the rue de Grenelle would lead to neglect, and indeed the decline of the following years seemed to confirm their worst fears.[31]

In 1963 Christian Fouchet, the education minister, created the collège d'enseignement secondaire, a general secondary school for young people aged 11 to 15, which grew to 2 million students in a decade. But the collèges secondaires still did not constitute a collège unique, for after the first two years of observation and orientation, they directed their students into classical, modern, or technical programs. In addition, 700,000 students attended the collèges techniques, essentially terminal institutions preparing students for vocational certificates. Technical education continued to be, in Faure's words, "l'enseignement de la dernière chance."[32]

The student revolts of 1968 reawakened old prejudices in industry against the universities as the makers of idle theorists and revolutionaries and reinforced the preference for professional schools, whether the *grandes écoles* or intermediate technical schools. The resignation of de Gaulle in 1969 and the presidencies of Georges Pompidou and Valéry Giscard d'Estaing during the 1970s brought a marked technocratic and probusiness orientation to government, which favored the professionalization of secondary and higher education. Suddenly l'enseignement technique was no longer the outcast of French education but its spoiled darling. Former education minister Edgar Faure stated flatly that "there is no such thing as a non-technical education." Pierre Billecocq, a highly placed educational official, pronounced that there was no longer to be any distinction between "une éducation meilleure et une éducation utilitaire. Nous sommes tous utilitaires."[33]

And so they were. While the *universitaires* sank increasingly into a mood of doom and gloom, victims of what Joseph Moody has called "the myth of the oppressed university"[34] the governments collaborated closely with business and industry to orient the educational system in a more professional direction. Between 1969 and 1971 close to a fifth of the national budget was spent on all levels of education, and the number of students in primary, secondary, and higher education, public and private, rose to 13.6 million, plus 620,000 teachers and around 200,000 administrators. As one educational official observed: "Henceforth one public employee in two is involved in the education of one Frenchman of four."[35] A good portion of much-increased funding was spent on l'enseignement technique. It cost the state 15,000–20,000 francs to provide a place in a collège technique as opposed to 7,000–10,000 in a regular college.[36]

Four laws passed in 1971 made continuing education "a national obligation" for young workers until the age of 18. Officials were concerned that so many young people (40 percent) left school without any vocational preparation and then worked in small firms and companies in which no professional training was provided.[37] The apprenticeship tax on employers was raised and the enforcement of the laws much improved. The laws also gave the technical education diplomas equality with those of the other branches of the education system and provided numerous "bridges" (*passerelles*) from one branch to another and into higher education. Most important, the new *baccalauréat de techniciens* came on stream in the early 1970s with 14 options. Orientation and counseling were much improved, and an effort was made to locate and develop the potentialities of each student. As a result, the number of young workers provided with professional certificates doubled, and the unemployment rate among them fell to 3.4 percent in 1975. The number of students in the lycées and collèges techniques rose from 720,000 to 820,000 in the early 1970s.[38]

In 1975 René Haby, the education minister, presented a law that converted the collèges techniques into lycées d'enseignement professionnel. These continued to prepare for the *certificat d'aptitude professionnelle* and the *brevet de l'enseignement technique* in preparation for going to work at age 16, but they also trained for the *baccalauréat de techniciens,* which led the way to the Instituts universitaires de technologie and thus into higher education. In the law of 1977 Haby also eliminated streaming in the collèges d'enseignement secondaire, which now became comprehensive schools for young people aged 11 to 15. Specialization in the lycées classiques, the lycées techniques, and the lycées d'enseignement professionnel did not occur until the ages of 16 to 18.[39]

By the end of the decade the number of students in technical schools and programs had reached 1.8 million, the same figure as in Germany.[40] The French had caught up after a century of trying. With so much growth, there were of course problems: bureaucracy, red tape, centralization, outdated programs and equipment, crises of morale, and declining standards, yet the number of young people leaving school without diplomas had fallen during the decade from 200,000 to 95,000 as a result of the system of "l'éducation permanente" that had been introduced in the Astier Law of 1919 and only fully realized 60 years later.[41]

The three most striking achievements of the 1970s, therefore, lay in the partial democratization of secondary education, the postponement of professional specialization until age 16, and some opening up of higher education to all social classes. The main obstacles to advancement for the working classes had always been the *baccalauréat,* the *classes préparatoires,* and the *concours* for the *grandes écoles.* The introduction of the *baccalauréat technique,* the *baccalauréat de techniciens,* the IUTs, and the *passerelles* since 1945 has overcome some but not all of the obstacles to higher education. The *passerelles* are especially important to the engineering schools, through which up to 40 percent of the students are recruited, as opposed to 20 percent in other fields.[42]

The main sources of inequality in French education today lie in the uneven quality of the lycées (students in key urban lycées do much better than students in rural ones), credential inflation, and the *grandes écoles*.[43] The Savary bill of 1983 sought to unify and democratize higher education by merging the *classes préparatoires* of the lycées into the universities, obliging the *grandes écoles* to recruit from the universities rather than from the advanced sections of the lycées. The purpose was to broaden recruitment, because, ironically, the average man has more access to the universities than to the preparatory sections of important urban lycées. The threat posed by the Savary bill to the preparatory classes galvanized the associations of secondary school professors and the alumni and professional organizations to strongly oppose the bill, which they succeeded in emasculating. But with their usual realism, the *grandes écoles* have begun to recruit the best students from all social classes, even if that sometimes means putting aside the once inviolable *concours*.

Technical Education and the Intellectual Community

Unlike classique and even moderne, l'enseignement technique never had many advocates in the intellectual community. During the nineteenth century no major thinker so much as noticed them, and the two or three second-level thinkers who did, such as Corbon and Le Play, criticized them for taking young people out of the home and workshop and placing them in the school. The Commerce group responded in kind with contempt for the mandarins of the intellectual world. The most articulate supporters of technical education were scientists and engineers such as Dupin, Morin, and Tresca. In the early twentieth century several thinkers attempted to define a *culture technique* in response to *culture générale*. One of the first was the philosopher Alain, defender of the republic of little men. Believing that clear ideas can only come from manual work ("penser sur l'objet"), he contrasted concrete knowledge with the abstractions of classical culture. He disliked "pensées qui ne coûtent rien," thoughts separated from life, lived on the margin of existence. "Penser sur l'objet" was, in Alain's mind, a way of thinking, speaking, and writing that always took real-life consequences into account.[44] Among others, poet Paul

Valéry, Catholic social thinker Emmanuel Mounier, the adventurer-novelist Antoine de Saint-Exupéry, and geographer-historian André Siegfried espoused ideas that appealed to the advocates of a *culture technique*. Siegfried, whose father Jules was a leading businessman from Le Havre and the minister of Commerce in 1893, favored a redefinition of Western culture on three bases: "un outillage, une technique, et une culture." "Society," he said, "lives by its technique; it survives and flourishes through its culture. The two are inseparable." [45]

Before World War II, however, it was difficult to find intellectuals who felt more than a general sympathy for the objectives of technical education. As late as 1929 M. Paul Arbousse-Bastide polled several hundred leading thinkers on the possibility of defining the "humanités techniques," and all but two replied that this was a contradiction in terms.[46] The *Compagnons de l'éducation nouvelle* favored popular education, but their projects for democratization and unification threatened the autonomy of the technical education division and its schools just as much as it did the privileges of classical studies.

During the interwar period a few individuals within the technical education division attempted to define a *culture technique*. Edmond Labbé, director of the division from 1920 to 1933, spoke of "les humanités techniques" but also emphasized vocationalism as a worthy goal: "Tout pour la profession, et par la profession." His successor, Hippolyte Luc (1933–1944), also sought to define a technical culture. In 1938 the law creating the centres d'apprentissage introduced "un complément de culture générale" in their programs. This constituted a significant departure from the vocationalism that had always characterized the middle-level technical schools.[47]

The experience of the war and resistance, the rapid economic expansion of the 1950s and 1960s, and the appearance of the Common Market saw the emergence of growing interest among intellectuals in defining a technical culture for the modern age. In addition to Langevin and Wallon, Frédéric Joliot-Curie, Georges Friedmann, Jean Fourastié, Henri Lefebvre, Pierre Naville, and Alfred Sauvy, scientists, economists, sociologists, and demographers, sought ways to humanize modern technology through reform of the educational system. Henri Wallon, a child psychologist and grandson of one of the founders of the Third Republic, demonstrated that practical intelligence contained many of the same elements as verbal intelligence, differing only as an "orientation of thought," not as something inherently less "noble" than verbal intelligence. He defined practical thought as "spatial intelligence" that seeks appropriate solutions to real problems (*une intelligence qui fait ses preuves*).[48]

Wallon believed that the improved counseling and orientation of students would help undo old prejudices against working people and manual labor. He favored a balanced intellectual formation (*polyvalence*) reconciling the old dualities between work and culture, general and technical education, and an "action éducatrice" in which the student played a role in the educational process. This dialectic involved what Wallon called *dépassement*—a breakthrough, a surmounting, a reconciliation of opposites. The old frontiers between technology and culture were to be overcome by a new technical humanism embodied in educational reform.

Georges Friedmann and Pierre Naville, both Marxists, predicted that under certain conditions man might come to dominate the machine. According to Friedmann, capitalist development had been characterized in the past by the degradation of work, "le travail en miettes." The breakneck advance of technology had made modern life ever more anarchical. The only hope, he concluded, was that automation constituted "a new stage in the dialectics of mechanization" affecting not only the techniques of production but also those of communication, transportation, and the use of leisure. "In all of these man is ceasing to be an actor and is becoming increasingly a sort of demiurge who conceives, initiates and supervises."

After visiting the United States, Friedmann concluded that opportunities for creativity and teamwork existed mainly in the advanced scientific industries. Research units of scientists, engineers, technicians, and skilled workers collaborated, taking an interest in the broader functioning of the factory. This process of "job enlargement" raised the possibility that work might become at once more democratic and social than in the previous stage of industrial development. But job enlargement necessitated important reforms in education, a lengthening of the period of instruction and an extension of continuing education to ensure the individual's adaptation to an economy in rapid transition. This meant enlarging the field of technical and scientific instruction and reorganizing all branches of education to account for the realities of the technological age. But if human values were to be preserved, it was just as important that the worker and citizen also acquire a general culture. This meant defining a new technical humanism to prepare the average man "to learn how to learn," to make intelligent choices, and to use his leisure time creatively. Friedmann predicted that segmentation would gradually give way in France to unified high schools providing a balance of general and professional training.[49]

Another rather disparate group emerged from the resistance and liberation with liberal and technocratic goals. These included Louis Armand, Gaston Berger, Jean Fourastié, and Alfred Sauvy, all of whom contributed to the journal *Technique, Art, Science* (*TAS*) which spoke for technical and professional education in France. They attacked the stagnation of the interwar years and the apparent return to the same patterns after World War II under the Fourth Republic. Sauvy, director of the Institut national des études démographiques and professor at the Collège de France, blamed stagnation on an aging population, the result of a century of low natality. This "turning away from life" by the French, and particularly the French bourgeoisie, accompanied a loss of faith in the future, an unwillingness to take risks and invest in the country. Despite its natural wealth and several generations of almost zero population growth, France was no richer in 1914 than a century earlier.

This led Sauvy to conclude that economic progress depended on "human capital" more than money. Despite total defeat and a prostrate postwar economy, West Germany had been able to rebuild and to absorb 12 million refugees from the east in just a few years after 1945. Sweden and Switzerland were also countries poor in resources but rich in human ingenuity, which had built themselves through hard work, careful management, and the adaptation of their educational institutions to economic needs. Sauvy predicted that the baby-boom of the postwar years, the first major population growth in France for a century, could lead to a more vigorous and youthful France. But the human potential involved in "la montée des jeunes" would be squandered if the school system was not transformed to meet their needs, and this included the creation of a first-rate system of technical and modern schools, as in Switzerland.[50]

Educational reformers of the postwar years, whether of Marxist, liberal, or Christian democratic persuasion, believed that the transformation of education was necessary to the adaptation of France to the modern industrial world. The purpose of education, they said, was no longer to teach a *culture générale* to an elite but rather to reach all young people, regardless of social origin, providing equal educational opportunity and a new curriculum based on a balance of manual, professional, and intellectual training (*polyvalence*). Such an open and flexible approach would enable young people to adapt to a changing technology over the course of their careers and provide them with a culture suitable to a leisure society.

By the 1960s and 1970s the knowledge revolution had created the conditions wherein one could hope that man might learn to dominate his techniques. The new continuous process plants, as opposed to the old mass production industries, raised the possibility that the worker might become a technician capable of managing machines and thus enjoying a degree of freedom from supervision, specialization, and routine. Once he had obtained a certain autonomy, he was more likely to see himself as a producer. An influential school of Marxist theorists, however, rejected the liberal, integrationist implications of this argument. Among others, Serge Mallet, in his *La Nouvelle Classe Ouvrière* (1963), predicted a rise in class conflict led by militant technical workers revolting against the older bureaucratic forms of management. After the revolutionary events of 1968, the main focus of educational thought shifted further away from integrationist arguments toward radical theories of conflict and change.[51]

While Mallet and other members of the left became increasingly disillusioned with the idea of education as a means of bringing about social change, the conservative governments of the 1970s saw in professional instruction a means of bringing education more closely into line with the requirements of the economy and perhaps also an excuse for limiting the swollen enrollments of the liberal arts. These governments introduced a number of reforms in secondary education, but they carefully avoided making any fundamental changes in the elitist nature of higher education and the *grandes écoles*.

Today most French students attend comprehensive high schools, but toward the end of the common program, at about age 16 or 17, they may opt for technical or vocational studies in the lycées techniques or in the lycées d'enseignement professionnel. The latter, introduced in the late 1970s, have regrouped all the vocational and professional programs. The technical and professional lycées open the way to higher technical education or (as is usually the case with the latter) to the IUTs. The rise of an integrated system of mass secondary technical education began in the 1950s, so it is probably too soon to evaluate the experiment. The problems of growth are familiar: disorientation and indiscipline associated especially with schools in urban areas and the concentration of children of foreign workers (now 10 percent of the schoolchildren in France) in professional programs. The disappearance of the technical education division in 1964 and the decision of *Technique, Art, Science,* the main forum for the exchange of ideas about "les humanités techniques," to cease publication in 1977 are an expression of a weakening of the old esprit de corps.[52] Surveys in the professional high schools have found that students are oriented toward their careers and small groups of friends and show little interest in intellectual and cultural matters. Only 3 percent approved of the liberal arts courses in the curriculum of their schools.[53]

The Association pour le développement de l'enseignement technique continues to promote a technical culture, but many of its 5,000 members are officials in the education administration, and their idealism seems to have waned. Technical education seems rather adrift in the huge education bureaucracy. There are far too many commissions, committees, orientation centers, courses, boards, diplomas, certificates, options, and programs for any but full-time professionals to keep track of. In such a situation a new *culture technique* has not evolved, nor has the lycée d'enseignement professionnel ceased to be more than "l'enseignement de la dernière chance" for many of France's young people.

The Ecoles d'Arts et Métiers, The Early Years, 1800–1880

Il faut aider à tout ce qui est utile,
Il faut attacher son nom à tout ce qui est bien.
Duke de La Rochefoucauld-Liancourt

In the following three chapters we shall trace the history of the Ecoles d'arts et métiers from 1800 to 1983, concentrating on the evolution of the programs and diplomas of the schools and the rise of the alumni association as the spokesman for all gadzarts. In chapters 8 and 9 we shall turn to the life of the schools and the rituals, traditions, and psychology of the students, and in the concluding two chapters we shall trace the social origins and careers of more than 2,000 graduates from 1810 to 1960.

Professional Instruction Under the Old Regime

Before the French Revolution the Catholic Church was responsible for most of the educational institutions in France. Religious orders such as the Christian Brothers ran the primary schools, and the Jesuits, Oratorians, and others directed the secondary institutions. The rather decadent universities were also partially run by the Church.[1] The state involved itself in education only to train the experts it needed, notably in the Ecole des ponts et chaussées and the Ecole des mines in Paris and in the military schools at Brest and Mézières. The idea of creating schools to train for the arts, trades, and industry evolved late because young people had traditionally learned their trades through apprenticeship. With the breakdown of the guild system during the eighteenth century and especially during the Revolution, the question arose as to how young people were to learn a trade. The Christian Brothers, founded by Jean-Baptiste de La Salle (1651–1719), had attempted to teach useful knowledge and oversee apprenticeships, but their schools had disappeared during the Revolution and were only slowly reestablished during the Empire.[2]

On the secular side, several cities in the north and northeast had established écoles de dessin and related écoles des arts décoratifs, notably at Besançon, Colmar, Dijon, Metz, Mulhouse, Nancy, Reims, and Saint-Etienne. In 1780 the Duke de La Rochefoucauld-Liancourt (1747–1827) drew up a plan for a school providing instruction in the three R's and the arts and crafts based on the factory schools he had observed in a trip to England. Six years later Louis XVI authorized the establishment of an "Ecole de Métiers" on the duke's farm ("La Montagne") near the village of Liancourt (Oise), allocating 10 sous per day per pupil. This institution opened in 1788 with 20 students, mainly the sons of soldiers from the duke's regiment. They were taught by lesser officers from the same regiment who worked from a program based on the three R's, military drill, and various crafts such as tailoring, shoemaking, carpentry, cabinetmaking, and locksmithing. By 1791 there were about 100 pupils, mostly the sons of soldiers.[3]

When Arthur Young visited the duke's estate in 1787, he discovered an orphanage for girls, a model farm, and several textile mills. On the eve of the Revolution a thousand people in the Liancourt area were employed in the duke's enterprises. The duke had introduced a comprehensive system of social security, including unemployment insurance, a retirement plan, and free medical care to protect the workers.[4] However, when he went into exile in 1792, much of his work fell into disarray. The school was moved into his château at Liancourt and gradually became a purely military institution for the sons of noncommissioned officers. In the meantime La Rochefoucauld traveled in Great Britain and the United States, publishing a long report on the model prison at Philadelphia in 1798.

The Consulate and Empire

When La Rochefoucauld returned to France after seven years in England and the United States, he realized more fully than ever the immense importance of the application of steam power to industry and the revolutionary transformations that this foretold for society. If France were to compete with Britain, she would have to introduce a system of public primary education and pay more attention to professional and technical education. The virtual disappearance of the apprenticeship system during the Revolution and the more complex technology of the new mechanical industries demanded that France train skilled workers and foremen as well as engineers and scientists.

La Rochefoucauld persuaded the first consul in 1799 to transfer the remnants of his old school at Liancourt to Compiègne as part of the "Prytanée français" of military schools located at Paris, Saint-Germain, and Saint-Cyr. Compiègne was established for about 400 sons of soldiers killed in war or deserving aid. Though in theory a school for apprentices teaching the three R's, mechanical drawing, and various crafts, it soon became a poor man's lycée teaching a mishmash of subjects to young men of all ages and backgrounds. Manual work had been abandoned as "too lowly for the children of many officers and as wounding the *amour propre* of the principal and other employees."[5]

La Rochefoucauld wanted to introduce a program of studies in the mechanical arts and persuaded his friend Chaptal, minister of the Interior, to speak to Bonaparte on the subject. The first consul agreed and decided to visit the school at Compiègne. When he asked one of the students what he wanted to do upon graduation, the student replied that he wanted to go into the army. Bonaparte observed that the "State makes a considerable sacrifice to bring up these young men, and when they have finished their studies, they are not, with the exception of the soldiers, of any use to their country. We will put an end to this; from now on we will train petty officers for industry."[6]

On returning to the capital Bonaparte set up a special commission, which included Monge, Berthollet, and Laplace, to develop a program of studies for the school and explore the possibility of establishing several schools of arts and trades. He responded to the commission's recommendation in the Decree of 6 ventôse an XI (25 February 1803), transforming the school at Compiègne into an Ecole d'arts et métiers to train skilled workers and foremen. He stated:

In the Nord I found excellent foremen ... but none who could make a measured drawing.... this is a real gap in industry. I want to fill it here. No Latin—that will be learned in the lycées just organized—but work in the various trades with the theory necessary to their progress. Here we shall train excellent foremen for industry.[7]

Beginning in 1804 the students were divided into three groups and given military ranks. The "artistes" over 13 years of age were formed into eight companies to study algebra, geometry, and industrial drawing and to work in various metal and woodworking shops and foundries. An elite of "aspirants" among the older boys took elements of advanced algebra, trigonometry, solid and descriptive geometry, mechanical physics, and chemistry and helped with discipline and instruction in the school. The younger pupils, called "commençants" and ranging in age from 7 to 13, studied the three R's, French grammar, and elementary drafting.[8]

In 1805 La Rochefoucauld inspected Compiègne at Chaptal's request and reported that he had discovered more than 50 students below the age of 8. Other students had arrived at age 15 or 16 without being able to read or write. There was no unity of studies and much dissension among teachers, professors, and the administration. Everything was in disorder; riots, theft, attacks on proctors, and "the most shocking vulgarity" were frequent. In his report the duke stressed the inadequacies of the château as a technical school and recommended transfer to another location. A year later Napoleon signed a decree (5 September 1806) that transferred the school to Châlons-sur-Marne in the Champagne region. On 8 December 1806 the students marched in military formation in winter weather to Châlons, 130 kilometers away, arriving on 2 January 1807.[9]

In the meantime an imperial decree of 1805 created a second school at Beaupréau in the west (Maine-et-Loire), which was not opened until 1811. This school was isolated, difficult to supply, and surrounded by hostile royalists. It was moved in 1815 to Angers during the White Terror. A third school was planned at Trèves but never built. Hence for many years Châlons was the only school that counted for much. As inspector general of the Ecoles d'arts et métiers from 1806 to 1823, La Rochefoucauld was, in the words of one official in the Ministry of the Interior, "the Czar of the schools."[10] Napoleon soon lost interest in the schools, and the Ministry of the Interior, under whose jurisdiction they fell until 1831, also showed little interest. In an age when the notion of the industrial sciences, and even of public primary education, was in its infancy, no precedent existed for the idea of public technical schools preparing skilled workers and technicians for industry. Indeed there was not much heavy industry in France at the beginning of the nineteenth century. The mechanical arts were associated with the vulgar, even dangerous working classes and therefore were not within the province of "honorable people." La Rochefoucauld's idea of a new pedagogy combining theory and practice, classroom and workshop, interested only a few liberal reformers. Moreover, the working classes, accustomed to apprenticeship and to learning on the job, were suspicious of school-trained workers and foremen. Consequently, skilled artisans and craftsmen were reluctant to send their sons to the schools.[11]

The old monastery of Châlons proved more suitable for conversion into a school of arts and trades than the château at Compiègne had been, but it was several years before the school operated on a regular basis. For lack of money and established routines, everything had either to be invented or improvised. This at least gave La Rochefoucauld the chance to experiment with his ideas on education. The program included introductory chemistry and physics, mechanics, applied geometry, elements of algebra, mechanical drawing, arithmetic, calligraphy, and French grammar. Elementary subjects were taught to the young or ill-prepared. There were workshops in fitting, blacksmithing, locksmithing, instrument making, carriage and furniture making, modeling, ornamental design, and file cutting.[12]

In his inspection tours of Châlons the duke sought to develop a pedagogy of technical education clearly separate from the programs of the lycées and integrating the teaching of theory and practice. He did not want to see classroom instruction and shop work placed in separate compartments. The goal was to produce workers and foremen who were at once skilled and intelligent:

The purpose of the school . . . is to place in the workshops of France trained and skilled artisans who are able to reason about their work instead of stupid workers who function mechanically—a type representative of the majority of workers today. In order to advance and perfect the mechanical arts, industrial instruction . . . must give youths all the knowledge necessary to becoming distinguished craftsmen and capable and enlightened foremen. . . . The instruction should be neither too formal nor directed to making scholars, nor should it merely teach a routine use of tools.[13]

Thus the duke was determined at Châlons "to lose no opportunity to turn a course into a lesson." Even the teaching of handwriting was to be divorced from "teaching words empty of meaning"; the pupils were to express the qualities of courage, charity, devotion, and honor or, alternatively, useful notions based on the natural sciences, mechanics, and industrial processes. The duke aimed at the integration of theory and practice in the various courses; for example, chemistry and physics professors could explain the composition of iron, its usages and properties, and methods of refining and manufacture, while in drafting, the student could draw plans of underground galleries of mines, the placement of beams, the design of water pumps and steam engines. "This enables a simultaneous advance in instruction at all levels and consequently the probability, even certitude, of success."[14]

Despite the duke's increasing intervention into the teaching at the school, he was not satisfied with the results. People did their jobs adequately, "but the contagious zeal for improvement, the desire to find ways to get things done well, to ensure success, is just not there. . . ."[15] In his reports on Châlons in 1806 and 1807, he stressed four obstacles to the development of the school: (1) the haphazard recruitment of students and their lack of aptitude for technical studies; (2) the absence of discipline; (3) the poor integration of the teaching of theory and practice, of classroom and shop, which he blamed on the faculty rather than the administration; and (4) the difficulty of placing graduates in industry.

No regular recruitment procedures or entrance examinations existed. The students were mainly the sons of soldiers, war veterans, or deceased soldiers. Napoleon and the minister of War sent boys of all ages, even preschoolers, to Châlons. It was as much a military orphanage as it was a school of arts and trades. Many pupils were scarcely literate on arrival, and few had any aptitude for technical studies; hence the courses were frequently beyond their comprehension and discipline suffered.

La Rochefoucauld blamed both the professors and students for the problems in the schools, especially for the failure to achieve unity of theory and practice in the programs. Professors at Châlons had their own methods and were unwilling to experiment with new approaches to learning. They interested themselves in their better students at the cost of the others: "The mathematics professor . . . teaches as if he were in a lycée and is satisfied if, of thirty students, two make it into the Ecole polytechnique."[16] The purpose of the Arts et Métiers was not to turn out academic elites but to prepare pupils for skilled positions in industry: "It is natural that the professors—and I speak particularly of those in mathematics, most of whom have recently graduated with distinction from the better schools—have a passion for their science and the desire to push ahead as far as they can in their teaching, but that which is desirable elsewhere is, in the school of arts and trades, an obstacle to good methods."[17]

La Rochefoucauld hoped to discover "good and zealous citizens among French savants who will give courses in the sciences as applied to industry."[18] But this was far from easy in a time when civil and mechanical engineering barely existed in France. The Ecole centrale, for example, was not established until 1829. Though France led the world in engineering theory in the early nineteenth century, the field was monopolized by the great state schools, the programs of which were oriented toward mathematics. Even England, the leading industrial power and the most advanced country in engineering practice, had not developed the teaching of the industrial sciences to any degree. Consequently, La Rochefoucauld had to work with professors who had little understanding of industrial education and who, especially the polytechniciens, disdained such applications as vulgar and mechanical. The professors were at once too theoretical and too specialized in their approach. There were also too many of them at Châlons: "We have five professors of grammar and only one or two students who can spell," commented the duke.[19]

True to the authoritarian reflex of the Napoleonic age, La Rochefoucauld advised the Ministry of the Interior that faculty and students alike would have to be brought more fully under the jurisdiction of the school administration. "The professors," he said, "will not easily yield to the idea of subordination, but they must be obliged to do so in order to assure that a hierarchical order is established." He concluded that "the principal [*proviseur*] is the head of administration and police; the professors must consider themselves as his subordinates, not his colleagues."[20] The principal became the bureaucrat par excellence. He was required to write a detailed monthly report to the minister of the Interior on every aspect of the school's life and economy. He was seconded by a dean of instruction (*chef d'enseignement*), the coordinator of teaching in classroom and shop, who spent much of his time procuring materials for the shops and foundries and selling the goods produced.[21] The proctors (*maîtres d'études*) were replaced by military guards (*surveillants*), who were ill-educated and crude and had no conception of academic goals.

In chapter 8, in which we discuss the life of the schools, we shall return to a more detailed evaluation of La Rochefoucauld's work. Suffice it to say for now that the military regime of *casernement* and stricter entrance requirements led to some short-term improvements in studies. However, very few graduates during the years of the Empire ever went into business and industry, and the few who can be traced were usually military men or government employees.[22] In 1811 Napoleon, increasingly in need of manpower for his military adventures, began to draft young men out of Châlons into the army. The carnage of the Russian campaign and heavy losses thereafter in the European campaign caused economic dislocation and undid the reforms in the schools. By 1814 La Rochefoucauld wrote to the minister of the Interior that the Arts et Métiers were becoming "charity schools, a place where parents can get rid of their children for three or four years."[23]

Nevertheless, the students remained attached to the memory of the swashbuckling days of the Empire. In 1814 the older students at Châlons, dressed in their school uniforms, attacked the invading Prussians on the plains of Champagne, thereby giving birth to a legend. During the Hundred Days the young men of Angers turned their school into a fortress and fought government troops rather than allow the tricolor to be replaced by the *fleur de lys*.[24] Such ardent patriotism in defense of the Empire did not make a very good impression on the Restoration regime, which the allies reimposed on France after Waterloo.

The Restoration, 1815–1830

The years from 1815 to 1823 were fairly quiet ones for the two schools, but the regime was suspicious of students' republican sympathies and La Rochefoucauld's liberalism. It could not forget that the duke had disdained to join the counterrevolution during the 1790s, preferring exile in the United States and Canada. Nevertheless, La Rochefoucauld was able to introduce some important reforms in 1817. He reduced the military character of the schools, though he kept the uniform. The students no longer ate, attended class, and went to bed at the beat of drums but rather did so to the ringing of bells, sounds more appropriate to the nineteenth century. He terminated the practice of putting students in the city jail for serious offenses, expelling them instead.[25]

The ordinance of 26 February 1817 improved the quality of programs and teaching, set age of entry at 13 to 16 years, and established the number of students at 400 for Châlons and 200 for Angers. The prefects were given the responsibility of recruiting students and establishing departmental examining committees to ensure minimum standards. However, they were not always assiduous in rounding up candidates for the examinations, who averaged only 2 to 4 per department. The Ministry of the Interior disposed of 50 places, which it usually gave to the children of court domestics. It also decided to admit some *pensionnaires* from the royal colleges to add some tone, but these boys, frequently rejects from the *collèges*, turned out to be the worst troublemakers of all.[26]

The schools gradually settled down during the early years of the

Restoration until the assassination of the *dauphin*, the Duc de Berri, in 1820. The regime then took a reactionary turn and became suspicious of the students' republicanism and of La Rochefoucauld's activities in various liberal educational and philanthropic societies.[27] Moreover, the Church wanted its two monasteries back so that it could convert them into *Grands Séminaires*. In 1823 the government announced that it was closing Châlons preparatory to transferring it to Toulouse. The city of Châlons protested, and Baron Charles Dupin defended the school in Parliament.[28] The decision was rescinded, but in 1824 La Rochefoucauld was retired as inspector general of the schools and indeed from all his semiofficial philanthropic activities. The principal at Châlons was replaced by the Vicomte de Boisset-Glassac, an ex-army officer and former member of *La Congrégation*, who insisted on detailed religious observances, played favorites among the students, and neglected technical education. This infuriated the students and provoked chronic conflict, which lasted until the end of the Restoration in 1830.[29] Indeed the long war between the students and the "Strass" (administration) began during this period. The turmoil of these years did result in some reforms, the raising of entrance standards and upgrading of programs in 1827 (and a few years later, in 1832, under the July Monarchy), which made it possible to gear the schools more closely to the nascent machine and metallurgical industries.

Prior to these reforms few graduates went to work in industry. La Rochefoucauld reported in 1819 that he had to keep 18 excellent students at Châlons for an extra year because "the stagnation of commerce has made it very difficult to find positions for foremen and workshop supervisors, functions which all of them are capable of filling."[30] The government decided to allocate 6,000 francs to subsidize the placement of the top 6 graduates each year, most of whom were assigned to government shipbuilding or arms companies at Indret, Rochefort, and Toulon or to firms working on state contracts.[31]

Before 1830 the Ecoles d'arts et métiers were in the curious position of preparing for an almost nonexistent heavy industry. Despite La Rochefoucauld's efforts, preindustrial patterns and an artisanal mentality persisted in the schools: no steam engines were installed until the 1840s; power to run the motors in the shops was provided by wind and water and sometimes by horses, casual laborers, and even by the students themselves.[32] It is not surprising that before the 1840s Châlons and Angers turned out more soldiers, government employees, and small craftsmen than industrial workers and foremen. The schools were probably established a generation too soon to be organized effectively for industry and were thus saddled with an artisanal mentality, a boarding school format, and a tradition of indiscipline that undermined the original ideas of La Rochefoucauld, Napoleon, and Chaptal.

The July Monarchy, 1830–1848

After the July Revolution of 1830, the government hesitated as to the future of Châlons and Angers. François Arago, the scientist and deputy, attacked the schools in the National Assembly: "The students are inadequate and incapable and have no alternative save to abandon careers in industry in order to seek jobs as tax collectors, customs officials or office clerks." [33] He submitted a bill in 1831 to replace the 2 Ecoles d'arts et métiers with 10 day schools to be established in key industrial cities. These were to teach young men aged 13 to 17 a combination of applied mathematics, drafting, and shop courses adapted to the local economy. Because his sound proposal for more technical schools was combined with an attack on the Arts et Métiers, defenders of the work of La Rochefoucauld were forced to oppose Arago's bill. Charles Dupin replied in the National Assembly: "Destroy, why do we always have to destroy? Certainly it is true that skill in execution can only be acquired through practical experience based on apprenticeship in a craft, but how many eighteen-year-olds possess such skills? The relevant theoretical knowledge can only be acquired in schools." [34] Arago's bill was narrowly defeated, but his idea of local technical schools was taken up for a time in the 1850s and again, with more success, in the 1880s.

In 1831 the Ecoles d'arts et métiers were placed under the authority of the Ministry of Commerce and Public Works. Adolphe Thiers, the minister, had been private secretary to the Duke de La Rochefoucauld in 1821 and was a strong supporter of his works. He hired capable directors for Angers and Châlons in Charles Dauban and J. A. Vincent and oversaw the reform of the schools in the law of 23 September 1832. This law reduced the length of studies from four to three years and set admissions at 100 new students annually, a total of 300 students per school—a system that lasted until the late 1920s. The minimum age of candidacy to the schools was raised from 13 to 14. Programs were consolidated, and the course content was adapted more directly to the burgeoning mechanical industries.

The upgrading of programs in 1832 accompanied the careful integration of practical and theoretical work. The students studied French composition and grammar, mechanics, physics, chemistry, advanced arithmetic, algebra up to quadratic equations, and applied and descriptive geometry, all with practical applications and problems. They learned how to draw machines in whole and in detail, dismantling complex pieces in order to analyze and draw each component. They designed and built machines and machine tools, which the schools sold. Students averaged five and one half hours in class and seven in the shops and foundries. Trades that could be learned by apprenticeship were eliminated, leaving four shops: machine construction, machine modeling, metal lathing, and founding.[35]

Backed by a sympathetic Ministry of Commerce, the new directors, Charles Dauban at Angers, a former mathematics professor at the Lycée Henri IV in Paris, and Jean-Antoine Vincent, a naval engineer and graduate of the Ecole polytechnique, oriented the schools toward the mechanical industries. Vincent, who became inspector general in 1837, had valuable contacts in industry and government. He arranged for the installation of the first steam engine at Châlons in 1839 and at Angers in 1848. The goal of the schools was increasingly to train skilled workers, foremen, and draftsmen who could be promoted to supervisory engineering positions as works and production managers.[36]

The reform of 1832 required all candidates to have completed at least one year of apprenticeship. The Société d'encouragement pour l'industrie nationale feared that parents of modest means could not afford the double cost of an apprenticeship and an education and that they would put their sons to work, depriving the schools of capable recruits. Several prefects correctly predicted that the number of applicants would decline.[37] But during the 1830s the quality of applicants gradually improved, enabling the schools to teach more science, mathematics, and engineering and spend less time on remedial subjects. In 1842 Vincent reported that students who had been apprenticed before coming to the school did better than those (about 20 percent) who had been excused.[38] The Guizot Law on Public Instruction of 1833 raised standards of elementary instruction, introduced teacher-training schools, and established higher primary schools for young people aged 12 to 14 living in cities of more than 5,000 inhabitants. In the decade after the passage of the law, the number of young people attending primary schools rose from 2 to 3 million.[39] This and, especially, the introduction of the higher primary schools made it possible for the Ecoles d'arts et métiers to recruit better-prepared students.

By the beginning of the 1840s the prefects were reporting strikingly improved interest in the Ecoles d'arts et métiers. In 1832, 300 candidates in all of France had presented themselves for the entrance examinations for 200 places; in 1840, 800 did so. The prefect of the Gard wrote in 1840 of "the ardor with which the best students from our primary schools compete for admission into the arts and trades school. This eagerness has become a kind of infatuation. Distance, financial sacrifices, nothing puts off the candidates, and yet most of them come from poor families who often have to borrow money for the trip to Nismes."[40] The prefect of the Isère wrote that in the examination of 1840 there were 13 candidates, "the largest number ever" and that "the examining commission was amazed by the ability of the top two candidates." The General Council of the Department of the Nord passed a resolution in 1839 calling for the establishment of a third arts and trades school in Lille. Similar representations were made by Angoulême, Bordeaux, Marseille, Montauban, Nîmes, and Toulouse.[41]

The demand for graduates was stimulated by advances in industry and technology. The Railway Act of 1842 inaugurated the construction of a national railroad network, which encouraged the development of the machine-construction and metallurgical industries. Initially the growth was not spectacular. France produced only 30 locomotives in 1842, and one of the main manufacturers, the Derosne-Cail Company, owed its success to the production of machines for the sugar beet industry. The entire French metallurgical industry at the end of the July Monarchy possessed 341 steam engines (up, however, from only 4, 30 years before).[42] French heavy industry still depended upon imported British technicians and machines. Gradually, however, the introduction of the railways, mechanized steamships, and overseas liners, the modernization of port facilities, the building of modern water and sewer systems and other public works in the cities, and the construction of big metal bridges, viaducts, and canals all stimulated the growth of heavy industry and the need for skilled industrial personnel.

In the southeast the demand for skilled personnel rose dramatically with the building of a big metallurgical and machine-construction complex at La Ciotat, near Marseille, which produced locomotives and other equipment for the railroad companies building the line between Paris, Lyon, and Marseille (the future **PLM**) and material for the burgeoning shipping industry of the region. These companies, the steamship lines, and the navy complained of competition for skilled personnel from the railroad companies; they could not find sufficient skilled workers and draftsmen from the region and were forced to import them at a high price from northern France and abroad.[43]

In 1838 the Ministry of Commerce sent a questionnaire to the prefects and general councils of the departments of southern France, requesting their choice for the location of a proposed third Ecole d'arts et métiers. The many replies and petitions to the Ministry of Commerce in the ensuing years stressed that the big state engineering schools produced "ingénieurs au petit pied" who had little knowledge of industrial techniques and were impatient to be promoted to top executive positions. The general council of the Department of the Bouches-du-Rhône noted that "scholarships to the arts and trades school [at Châlons], once so difficult to fill, are now eagerly sought after."[44] The competition among the cities of Aix-Marseille, Nîmes, and Toulouse for the new school was spirited. Fifteen of the southern departments polled by the Ministry of Commerce favored Toulouse because of its central position between the cities of Bordeaux and Marseille "entre les deux mers."[45] Toulouse made a generous offer of 500,000 francs, and the Department of the Haute-Garonne granted 100,000 more. However, there was no building in Toulouse suitable for cheap and rapid conversion; a new school would cost 1.7 million francs and would take three years to build.

Four departments voted for Nîmes (Gard), one for Montpellier (Hérault), and one for Aix-en-Provence (Bouches-du-Rhône). Nîmes was favored because of its proximity to Marseille and Lyon, but its industry (mainly silk), its distance from seaports, and its relatively modest offer (216,000 francs) disqualified it. In 1840 the Department of the Bouches-du-Rhône, joined by the Var, of which Toulon is the main city, launched a determined campaign to win the school for Aix, located a few kilometers from Marseille. The two departments and the city of Marseille offered 212,000 francs, and the city of Aix ceded a monastery that had already been converted into a collège during the Restoration. Capable of housing 500 people in spacious buildings and grounds, already serviced, it was projected to cost a relatively modest 250,000 francs for conversion into an Ecole d'arts et métiers, a figure that the departments and municipalities could meet by slightly enlarging their grants.[46]

The minister of Commerce, Cunin-Gridaine, was a close associate of many industrialists, notably Ernest Gouin and Paulin and Joseph-Léon Talabot. The latter two were major backers of the PLM.[47] Adolphe Thiers—deputy from Marseille, former secretary to La Rochefoucauld, and minister of Commerce at the time of the reform of 1832—led the other seven deputies from the area in a parliamentary campaign to obtain the school for Aix-en-Provence.[48] Soon all obstacles had been cleared away; in the spring of 1843 Parliament voted to create a third Ecole d'arts et métiers at Aix, which opened its doors in October 1843. Aix provided skilled personnel for the machine and metallurgical industries of southern France, the PLM, and for the navy and merchant marine. By the 1860s more than half of French naval technical officers came from the Ecoles d'arts et métiers, mainly from Aix.

Though employment prospects improved considerably during the 1840s, graduates sometimes had difficulty finding jobs. The three schools were lost in small provincial cities, and aside from the Ministry of Commerce and the inspector general, there was little liaison between them and even less among the graduates. Efforts in 1823 and 1839 to establish an alumni association had failed. Several informal groups of gadzarts had sprung up in major industrial cities, notably Paris and Marseille. In June 1846 the Paris group again attempted to obtain permission to organize an alumni society. When a committee met with M. Gabriel Delessert, prefect of police, to obtain authorization under the association laws, the latter observed: "Je comprends. C'est de l'égoïsme en grand que vous voulez faire."[49] But authorization came anyway, and an inaugural banquet was held on 4 July 1847 with 130 participants.

Although the association sometimes acted as a spokesman for the interests of the schools and their graduates, it saw itself mainly as a friendly and mutual aid society providing services and contacts to its members. Its employment office assisted in the placement of graduates and in the advancement of those already working in industry. The association had some ties with the Société des ingénieurs civils de France, founded in 1848 mainly by centraliens, and various other industrial organizations. But not until some years later did it become a semiprofessional organization lobbying to improve the status and prestige of the Ecoles d'arts et métiers in the world of education and industry.

The Second Republic, 1848–1851

The alumni association had been organized little more than a year when it found itself defending the schools from their critics during the reaction that followed the revolutionary events of 1848. The events of February to June of that year, in which the students and a few graduates played a small but perceptible role, confirmed the bad reputation of gadzarts in conservative circles. Once again the schools were criticized for wasting the taxpayers' money, for turning out "soldiers, artists, and dancing masters," unemployable and potentially revolutionary malcontents.

The result was a serious movement to suppress the schools in the budget discussions for 1850 and 1851 in the National Assembly. In the sitting of 26 April 1850, article 10 of the budget reduced the grant to the Arts et Métiers by 14,000 francs (from 1,165,000 francs to 1,151,000). An amendment of M. Corne, a deputy from Lille who was favorable to the schools, advocated a return to the original credit, while M. Raudot, an opponent, proposed closing the schools altogether. Raudot, backed by Berryer, the reporter of the budget, argued that they taught too much theory and turned out mediocre workers. He admitted that a few first-rate engineers and supervisors had emerged but claimed that these were exceptional cases: "For the majority of young men, the instruction is too advanced and is a source of bitter disappointment. More than nine-tenths leave the schools unable to make a living.... These young men become malcontents accusing a society that has raised their hopes without being in a position to fulfill them." Raudot argued that the government should get out of professional education, leaving it to apprenticeship, to factory schools, and to local governments.[50]

The alumni association executive quickly got in touch with the main employers of gadzarts. Just before the debate of 26 April, a petition reached the minister of Commerce, J. B. Dumas, signed by leading industrialists (Cail, Clapeyron, Flachat, and others) who declared that the schools produced men of "genuine merit as skilled workers, mechanics, engine drivers, draftsmen and industrial engineers." They concluded that "the Ecoles d'arts et métiers are of an incontestable utility; if they should cease to exist it would be necessary to recreate them.... Indeed if by some foolish stroke these schools were suppressed, national industry would receive an indescribably serious blow."[51] M. Vernilhac, director of the Mediterranean Steamship Company, wrote: "The Ecoles d'arts et métiers seem to me to be the best, and really the only, source of naval mechanics and technicians who are qualified to serve in our steam-powered fleet. Under the able direction of M. Vincent, a naval engineer, they are furnishing us every day with first-rate men."[52]

The alumni association also did a quick survey of all the graduates they could account for, tracking down 1,343, or about one third of living graduates. Of these, two thirds (907) worked in industry or the railways in positions ranging from skilled workers, foremen, and draftsmen to factory managers, company directors, and successful industrialists.[53] The association's findings proved the falsity of Raudot's charge that the "great majority of graduates are incapable of making a living."

Such evidence, backed by the testimony of industrialists, provided sympathetic deputies such as Hippolyte Corne with the necessary ammunition for a spirited defense of the Arts et Métiers. Corne's department, the Nord, had petitioned the Ministry of Commerce for years to establish a fourth school in Lille, and it maintained several scholarships at Châlons. Corne emphasized that a sound knowledge of industrial drawing, applied geometry, and mechanics could not be learned on the job. He said, "In France we have three schools for a working population of thirty million, while we have three hundred lycées and collèges teaching the classics to an infinitely smaller population of five or six million."[54]

The assembly voted the slightly reduced budget of 1,151,000 francs for 1850 and soon after moved to consider the budget of 1851 and the question of the schools' suppression. By this time the minister of Commerce, noted chemist J. B. Dumas, who had initially temporized, came forth in support of the schools. Charles Dupin and Arthur Morin, director of the Conservatoire des arts et métiers of Paris, also vigorously defended them. In response to Raudot and Berryer's criticisms that the Ecoles d'arts et métiers were too theoretical in orientation and failed to train skilled workers, Morin argued that this was in fact their virtue. A school graduating 300 workers per year would have little impact on the mechanical industries, but one producing 300 foremen and technical supervisors would make an important contribution to industrial development. The gadzarts were rapidly getting ahead in industry while contributing to economic growth. It was, said Morin, the classical lycées and collèges that were turning their students away from productive enterprise. The *universitaires*, he concluded, were useless bureaucrats, "quite incapable of training professors for the applied sciences and certainly not for the mechanical arts." If such men were not trained by the Ministry of Commerce, they would not be trained at all.[55]

These arguments won the day, and the National Assembly, by a vote of 381 to 210, rejected the cuts in the schools' budget for 1851. The Ecoles d'arts et métiers survived and were never again seriously threatened. However, the higher primary schools from which they recruited were not mentioned in the Falloux Law on public education of 1850 and survived only in a few industrial cities as municipal or private institutions (see chapter 3). As a result the Arts et Métiers had some difficulty obtaining qualified students, and their development as secondary technical schools was hindered.

The Second Empire, 1852–1870

The loss of many of the higher primary schools and the lack of professional schools disillusioned many proponents of technical education with the *Université*, but they did not always agree among themselves as to what they wanted. The 1850s and 1860s saw a vigorous debate between those favoring training on the job and those advocating the organization of a national system of professional schools under the Ministry of Industry and Commerce, a *Université du Travail* as they called it. The former group was led by Anthime Corbon, Frédéric Le Play, and by some industrialists. The latter group was composed of Morin and Tresca and their associates in the Conservatoire des arts et métiers, by some gadzarts, notably André Guettier (Châlons, 1832), and by the Société des ingénieurs civils, which endorsed the idea of the organization of professional education shortly after its founding in 1848. These groups, gathered loosely around the Ministry of Industry and Commerce, defended the interests of technical instruction. They accused Corbon and Le Play and their followers of having an old-fashioned and artisanal approach. They noted that the substitution of metals for wood, the conversion to steam power, and the building of a national railway system encouraged standardization, mechanization, and the concentration of heavy industry, which in turn increased the demand for school-trained industrial specialists. They pointed out that by the 1850s imported British technicians were being replaced by French ones, educated mainly in the Ecoles d'arts et métiers, and that France had become an exporter of metals, machinery, and locomotives.[56] As the structure of industry became more complex and manpower demands more varied, industrialists began to look to technical schools as the source of skilled personnel. They had no confidence in the traditional *Université* to provide such training and increasingly favored the idea of an expanded school system under the Ministry of Industry and Commerce separate from the schools under the Ministry of Public Instruction.

The first gadzarts to develop a theory of industrial education was the manufacturer André Guettier, who advocated a *Université des Arts et Métiers* separate from but competing with the traditional *Université*.[57] He wanted the state to clearly define the title of engineer and then take measures to protect it from phony claimants and "doers" who obtained the title through job promotion. The engineering diploma should receive the same protection as those of architecture, law, and medicine. In the new *Université des Arts et Métiers* there would be technical "baccalauréats and doctorates of industry" as well as engineering credentials. The key difference was that the graduates of all engineering faculties and schools, including Centrale and Polytechnique, would receive only the title "aspirant engineer." Full award of the diploma would come only when the individual had proven himself on the job. Graduates of the Ecoles d'arts et métiers would obtain the same title but only after a fourth year of study in the advanced sciences, languages, and the arts had given them a technical culture.

Most gadzarts took a position somewhere between those of Corbon and Guettier. While some agreed with the idea of a *Université du Travail*, they were unhappy about Guettier's plan to prevent "doers" from winning the engineering title through job promotion, because so many of them had acquired the title in precisely that way. They saw themselves as men of action, as self-made men, and did not want technical education to become as formalized as the other branches of the education system. They feared that the Arts et Métiers might become just another institution defending its *position acquise* in the dog-eat-dog world of the educational bureaucracy.

Nicolas Cadiat (Châlons, 1820)—inventor, structural engineer, and industrialist—opposed Guettier's plan, arguing that industry was the only branch of French life unafflicted with a plethora of certificates and legal privileges.[58] After Cadiat's premature death in 1857, his position as spokesman for the free enterprise position in the alumni association was taken up by two of the most successful industrialists produced by the Arts et Métiers during the nineteenth century: Hippolyte Petin (Châlons, 1828), head of one of France's largest iron and steel companies, and Hippolyte Fontaine (Châlons, 1848), an associate of Gramme in demonstrating the reversibility of the dynamo.[59] Petin, the son of a blacksmith, and Fontaine, the son of a carpenter, opposed any elevation of the programs and diplomas of the Arts et Métiers that would convert them into imitation *grandes écoles*, denying access to the laboring classes. Their position prevailed in the alumni association through the 1870s. As a result the association executive did not push very hard for the upgrading of programs and diplomas for fear of placing them beyond the reach of the working classes.

The first reform of the schools undertaken by the Ministry of Commerce since the law of 1832 took place on 30 December 1865 with little association input. The law introduced upgraded courses in chemistry and physics and modernized the shops and equipment. It revised the official goals of the schools to produce "shop supervisors and workers trained in the enlightened practice of the industrial arts."[60] Many members were unhappy with a definition that continued to see graduates as workers and failed to recognize their achievements in industry.

In his retirement speech after 16 years as president of the association, industrialist and former deputy Henri Flaud (Angers, 1830) commented in 1866 that "the schools actually produce educated industrialists and not soldiers and petty officers of manual work." In reference to the law of 1865, he asked: "Why state in advance that the graduate is only going to become a skilled worker? No one has the right to assign to a young man who has just graduated from the schools a set place in the industrial ladder.... Why not give to these young men the right to become engineers if they are capable of it, [especially as] the majority of our machine construction factories are directed by graduates of the Ecoles d'arts et métiers." He concluded: "Once the public has accepted that the son of a worker or laborer can become a marshal of France by his courage and talent it will hardly deny him the right to advance himself through his own work to the highest positions in industry."[61]

Flaud's remarks typified the mixed feelings of gadzarts about social and professional advancement. Anxious to maintain their independence and devoted to free enterprise, they were also ambitious to improve the school's image and their own professional standing. They resented the lack of recognition of their achievements and bitterly criticized the privileges and monopolies of the big engineering schools of Paris. At the same time they were tempted to imitate the example of the Ecole polytechnique as a highly successful educational lobby. Thus the alumni association was gradually drawn into politics. In

1860 the Second Empire legally recognized it as being "of public utility," which meant that it could receive donations and legacies. A subsequent gift from the La Rochefoucauld family enabled it to considerably extend its operations, and its membership rose from 326 in 1860 to 1,220 a decade later.

The influence of highly placed businessmen on the executive committee of the association facilitated contacts with other industrial and professional associations and with the Ministry of Industry and Commerce. In 1873 the association obtained five seats on the newly established Conseil supérieur de l'enseignement technique. The following year the council set up a committee, chaired by Hippolyte Fontaine and composed mainly of members of the alumni association, to recommend reforms for the Ecoles d'arts et métiers. Dominated by the old guard, it decided against any major changes for fear of turning the schools into "branches of the Ecole centrale," engineering schools out of touch with the working classes.[62] But as we shall see in the following chapter, this would be the last victory for the traditionalists over those wishing to upgrade the schools' prestige.

The alumni association entered politics mainly to defend the interests of the schools and their graduates. It especially opposed the monopoly of the polytechniciens over engineering positions in the corps of roads and bridges. The corps had the power to approve all transportation projects in which the state had an interest (railway networks, roads, bridges, canals, port facilities); hence they could make life difficult for entrepreneurs and civil engineers working on such projects. In addition, the corps employed middle-level technicians called *conducteurs des ponts et chaussées* to supervise construction projects. In the early years, before industry began to offer abundant opportunities, many gadzarts entered the department of roads and bridges and were confined to the post of *conducteur*. Despite the considerable competence they acquired on the job, *conducteurs* could not be promoted to engineer because of the legal monopoly that polytechniciens held over the post.[63]

In his study of the *conducteurs* before 1850, John Weiss found that 27 of 270 had been trained in an Ecole d'arts et métiers. As a group the gadzarts moved up the hierarchy from *conducteur* third class to first class and then to *conducteur principal* more rapidly than their colleagues and were more likely to be allowed "to perform the functions of engineer" on an interim basis. When the qualifications for access to engineering positions in the corps of roads and bridges were eventually relaxed somewhat, the first to pass the difficult examinations, in 1868, was Jean Caillé, a graduate of Châlons in 1851.[64]

Why did gadzarts accept jobs in a corps that held out little or no hope for promotion to the higher ranks? Most entered the corps during the first half of the century, when positions in industry were still not easily available, especially before 1840. Few did so thereafter. Some may have taken jobs to be near home, because of employment security and, especially, because of the retirement pensions associated with the public services and seldom provided by private industry. Most apparently stayed in the corps for only a certain period of time before moving on to posts in industry or the railways. As Weiss noted and as we will see later, they were a restless lot, especially in the earlier years of their careers.[65]

In France the state has always played a central role in the creation of elites, mainly by its control over the *grandes écoles* and its monopoly over certification examinations for which these schools prepared. Despite their dislike of the Ecole polytechnique and other *grandes écoles*, gadzarts were inevitably tempted to imitate these successful vested interests and to encourage their alumni association to use its contacts to obtain from the Ministry of Commerce an upgrading of programs and diplomas, even at the cost of sacrificing some of the popular and democratic aspects of the schools. From 1850 to 1880 the free enterprisers in the association were strong enough to contain their ambitious colleagues, but with the coming of a parliamentary republic in the 1880s intent on restructuring the French educational system, the position of the political group was much strengthened.

Despite their differences, gadzarts were united in their dislike of the old *Université*, which had made only one concession to the modern world since the First Empire, the watered-down l'enseignement spécial of Victor Duruy in the 1860s. The issue of capital and labor was, in their minds, a minor one in contrast to the deeper truth that both bosses and workers were victims of the babbling bureaucrats of the public education system with their disdain for the nation's productive forces. Translated into political rhetoric, such beliefs as these united gadzarts, disguised their differences, and brought them closer to members of other industrial and alumni associations linked to the Ministry of Commerce. Under the Empire and the Moral Order, a technical education lobby was taking shape.

The Ecoles d'Arts et Métiers and The Third Republic, 1880–1914

La France ne doit pas s'endormir, elle a à ses portes des concurrents redoutables dans l'ordre du travail. Sur le champ de bataille industriel comme sur l'autre les nations peuvent être surprises et périr en perdant des supériorités incontestables. Relever l'atelier à l'heure qu'il est, c'est donc relever la patrie.

Jules Ferry, opening of the Ecole nationale professionnelle of Vierzon.

In 1880 the Ecoles d'arts et métiers celebrated their centennial. From the point of view of historical accuracy, the celebration took place years too soon, for La Rochefoucauld had not actually opened his school at Liancourt until 1788. Moreover, it is quite arguable that the first true school of arts and trades was created not by La Rochefoucauld but by Napoleon in 1803, when he converted the military institution at Compiègne into an Ecole d'arts et métiers. But the two Bonapartes had destroyed the first two republics, hardly a strong recommendation in the early years of the Third Republic, while La Rochefoucauld had been a man of impeccable liberal credentials, enjoying somewhat the same reputation in liberal circles in France as did Thomas Jefferson and Benjamin Franklin in the United States. He provided an aura of liberal gentility that nicely balanced the reputation of the Arts et Métiers as schools for mechanics. Thus the alumni association sought to associate the schools whenever possible with their founder.[1]

The duke's apotheosis took place on 8 August 1880 on his old estate at Liancourt. The association chartered a train to take 550 *convives* from Paris. Accompanied by fulsome speeches and torrents of gratitude ("He has made us all that we are"), a large monument was raised to the founder's memory near a statue of him that the association had erected in 1861. An afternoon banquet at Liancourt was followed by an even more sumptuous one that evening in Paris, attended by 600 guests, including prominent politicians, officials of several ministries, and officers of various educational, industrial, and professional associations. In both banquets, "The representatives of the press were seated at the table of honor. . . . Never have journalists been surrounded with more care and attention than they were here."[2]

The celebration's timing could not have been better for a well-orchestrated campaign in favor of the schools. The new parliamentary republic was making primary education compulsory and free and re-creating a system of higher primary schools. These included a number of professional schools located in industrial cities and called Ecoles d'apprentissage and Ecoles nationales professionnelles. The latter, at Vierzon, Voiron, Armentières, and Nantes, were regional boarding schools designed to train skilled workers and foremen. Their programs and goals overlapped with those of the Ecoles d'arts et métiers. The Ministry of Public Instruction was apparently moving determinedly into the field of intermediate technical education, threatening the small system under the Ministry of Commerce.

For the Ecoles d'arts et métiers, the new professional schools provided both an opportunity and a threat. If the Arts et Métiers were promoted to the secondary level, they could recruit qualified students from these schools and offer more advanced programs, preparing their most able students to enter the Ecole centrale. However, if the higher primary professional schools could not be properly integrated with those under Commerce, the Ecoles d'arts et métiers might be submerged in "a sea of new creations," becoming little more than arts and trades schools.[3]

Many members concluded that the alumni association must become at once more democratic and more political, a cross between a professional organization and an educational lobby. The appearance of a parliamentary regime, a growing educational bureaucracy, and an increasingly centralized heavy industry favored the formation of professional associations having close ties with educational institutions. By the 1880s the Ecoles d'arts et métiers had become the principal supplier of mechanical engineers in France. Even though the schools were not formally allowed to grant a diploma until 1907, many graduates obtained the title of mechanical engineer through job promotion. Yet the schools' prospectus still defined the schools as training workers, foremen, and shop supervisors "skilled in the mechanical arts." The gap between the official prospectus and the achievements of graduates misled public opinion and led to frequent criticism from the left and the right that, as boarding schools, they spent too much public money training too few skilled workers and foremen.

The alumni association sought to correct such misunderstandings and promote the schools' interests by gaining greater control over school policy and direction, upgrading programs and diplomas, and creating an esprit de corps among students and graduates. The association portrayed gadzarts as producers struggling to build the economy and strengthen France against the German menace. It deplored "useless classical education," the privileges of the *grandes écoles*, and the evils of bureaucracy as sapping the national fiber. To this end it allied with other educational and professional associations grouped around the Ministry of Commerce. Beginning in the 1880s these groups formed a technical education lobby to support the ministry's efforts to wrest control over the Ecoles nationales professionnelles and the Ecoles pratiques de commerce et d'industrie (the former écoles d'apprentissage) from the Ministry of Public Instruction. Once the Ministry of Commerce obtained authority over these schools, it could organize a viable *Université du Travail*, with the higher primary professional schools at the base, the Ecoles d'arts et métiers at the secondary level, and the Ecole centrale on top.[4]

In the meantime, however, the Ecoles d'arts et métiers were vulnerable. For two generations they had existed in isolation, without any real competition from other schools, and had changed very little during these years. They recruited students from various municipal and private higher primary professional schools and required no diploma for admission. Their own *brevet* conferred no legal rights or privileges, and their programs went no further than elementary mechanics, descriptive geometry, and advanced algebra. Graduates had to scramble for jobs, often starting out near the bottom of the industrial ladder, where they frequently encountered hostility from workers and foremen who feared competition from school-trained workers.[5] Before the 1840s many graduates did not go into industry, and before 1880, 40 percent never obtained the *brevet*. Buried in small provincial cities, the schools were located some distance from industry, scientific institutions, and research centers. The regime of *clôture* and *casernement* further isolated them and provoked no end of troubles, the worst of which were exploited by the press.[6]

In 1880 the association was galvanized into action by renewed attacks on the Ecoles d'arts et métiers by Anthime Corbon in the republican journal *Le Siècle*. Such attacks were a rude blow to the gadzarts, many of whom were the sons of workers and republican in sympathy. Corbon, a former worker and a senator, stated that the Arts et Métiers produced only "hommes de bureau" and "mandarins de second ordre." The *Siècle* ran an editorial criticizing the low level of studies and shop work and claiming that the average student learned just enough to escape from doing manual work in the factory. For the price of a boarding school education, many more young workers could be instructed in day schools such as the écoles d'apprentissage.[7]

Coming in a republican newspaper on the eve of important changes in French education, Corbon's attacks spurred the executive of the alumni association, hitherto hesitant, to publicly defend the schools and also to seek a statute clearly establishing the Ecoles d'arts et métiers at a level above the new higher primary professional schools. The association president, Lucien Arbel (Aix, 1843), was a prominent metallurgist and arms manufacturer who had been a deputy and then senator from the Loire from 1871 to 1888. A conservative, self-made man, the son of a day laborer, he was reluctant to involve the association in politics, but he now found himself leading a campaign against *Le Siècle* and Corbon. Though he succeeded in obtaining a partial retraction from the newspaper concerning the graduates' career achievements, the newspaper insisted that such achievements resulted from students' native abilities rather than from the education they had received.[8] More important, Arbel obtained from the minister of Commerce five seats for the alumni association in the Conseil supérieur de l'enseignement technique. The association delegates (Lucien Arbel, Auguste Albaret, Antoine Léon, Java Mignon, and Louis Martin), all prominent men, set to work in the council to obtain a new law governing the Ecoles d'arts et métiers.[9]

In the meantime the association was being reorganized in order to play a more important public role in the new republic. It had grown from 1,000 to 2,000 members between 1870 and 1880 but was still a good deal smaller than the alumni associations of the Ecole centrale (3,000) and the Ecole polytechnique (4,500).[10] It was also overly centralized in Paris; for example, the executive committee was elected by the annual general assembly in Paris, attended by a hundred or so graduates, most of them wealthy, retired industrialists who had the time and leisure to look after association affairs. The conservative Arbel, a former senator closely associated with the politics of the Moral Order, was clearly not the man to transform the association. A younger man of impeccable republican credentials and political contacts was needed, preferably one who could provide the association with a social theory and an attractive public image. This man was Denis Poulot (Châlons, 1847).

Poulot was one of the rare graduates who was at once a successful businessman and articulate social commentator. The son of a small manufacturer at Gray (Haute-Saône), Poulot finished his studies at Châlons in 1850, worked as a fitter in the machine shop of his brother, and then became the foreman in a large machine-construction company. At the age of 25 he established a successful iron works at Belleville, sold it, and in 1872 founded a machine-tool company in the eleventh arrondissement of Paris. This firm having prospered, Poulot was elected mayor in 1872 of the eleventh arrondissement, the most industrial and populous in Paris, a position he held until 1882, when he became president of the alumni association of the Arts et Métiers.

Poulot had become known in France a decade earlier as a result of his book *Le Sublime* (1872).[11] A realistic account of the degradation of work in the factory system—the alcoholism, poverty, and decline of moral values—it formed the model for Zola's *L'Assommoire*.[12] Poulot, a factory owner, was certainly no socialist predicting the inevitability of class conflict. He believed that industrial growth would benefit everyone in the long run but that means had to be found in the interim to humanize work and heal the rifts within the community of production. A fervant republican and anticlerical, "un bourgeois démocrate," as he called himself, he saw in the "République une et indivisible" the best way of achieving social harmony.[13]

Poulot argued that the professional school was the only place where the young worker could acquire the necessary science and skills, along with a general education and moral and civic instruction. He advocated the introduction of compulsory shop work in the elementary schools and wrote textbooks on manual instruction. He saw a democratic primary system oriented toward the world of work as the base of what he called the *Faculté du Travail*. This plan was similar to the *Université du Travail* of the Second Empire except that it was to exist within a reformed *Université*, not outside and in competition with it. The Republic alone, one and indivisible in education, could guarantee a genuine opening up of careers to the talents of young workers and the end to "the monstrous system of feudal privilege" that had hitherto ruled in France.[14]

In reply to the criticism of Corbon and others that training workers in schools was a waste of time and money, Poulot said that modern capitalist society gradually opened up all kinds of intermediary positions requiring an increasing amount of education and that the socialist division of society into workers and bourgeois was an oversimplification. To prove his point, he studied his own *promotion* of 1847 at Châlons. He found that most were the sons of workers, farmers, shopkeepers, and small manufacturers. Thirty years after graduation, 40 percent had become owners or directors; only 11 percent were still employed as skilled workers, mechanics, engine drivers, and draftsmen (see chapter 10). They had reached high positions through hard work and with "the intellectual capital amassed solely at the Arts et Métiers."[15] His colleague Ernest Trouillet studied his own *promotion* (1847–1850) at Angers and reached very similar conclusions.[16]

The results of the two studies gave Poulot a theme that he used as a defense against Corbon's attacks and as a rallying cry to all gadzarts. At the annual banquet in Paris in 1881, he uttered the words that were to be cited for years to come:

Ecoutez bien ceci, jeunes camarades: le polytechnicien, de par une Société financière ou autre, *administre* une industrie, un établissement, un chantier. L'élève de l'Ecole centrale *achète* une usine. Le Gadzart, lui, *crée, fonde* un atelier.

Savez-vous pourquoi?—parce qui'il est entré dans la boîte à fumée.[17]

The *boîte à fumée* metaphor was seen as both a profound truth and an uproariously funny play on words having to do with the smoke of the workshop and the stench of the barnyard. Unlike the polytechnicien, who graduated with an engineering credential and a guaranteed place in the state corps, and the centralien, who graduated with an engineering credential and who had family connections that ensured him a good position in business and industry, the gadzarts had "neither title, place nor fortune." He had nothing to sell but his work. This is why he had to enter the *boîte à fumée*, if only temporarily, and work his way up. It was just this spirit of enterprise and tenacity that France needed in its struggle with foreign competitors. The gadzarts did not need to feel inferior; he was France's finest.

For Poulot the real enemy was *fonctionnarisme*, parasitical bureaucracy fed by privileged special schools and useless classical education. Capital and labor should cease to fight each other, for the real enemies of all producers were the "faux littéraires," the "déclassé intellectuals," and the bureaucrats who transformed producers into victims. He told the workers:

It is not Mr. Capital that you should be knocking; it is the social leprosy that comes from the four principles of saber, cassock, toga and red tape. Until society rids itself of their domination, nothing will change. Neither your hatred nor your sacrifices will do any good. So leave Mr. Capital alone. For the moment there is nothing much he can do. Go to work! When you have proved yourself he will give you his daughter in marriage.[18]

In his campaign for the presidency of the alumni association in 1881, Poulot argued for decentralization, greater political activity in favor of the schools, and a more liberal social policy. The opposition, led by César-Auguste Trotabas (Aix, 1844), a retired naval lieutenant, feared that distinguished men would not present themselves for office if the executive were to be elected democratically by mail ballot and that the association would become a special-interest group at the cost of its other activities. Poulot's chief supporter, M. Besnard, replied, "Most of us are the sons of workers. . . . We have a debt of honor to concern ourselves with the situation of other workers' sons." Poulot was elected by 145 of 170 votes cast.[19] Clearly he had a mandate for change.

Poulot named an education commission to examine the reform of the Ecoles d'arts et métiers. Reporting in 1884, it recommended the introduction of more advanced courses and the modernization of the workshops. It called for the formation of a board of governors (*conseil de perfectionnement*) to oversee programs, teaching, and discipline in the schools.[20] These recommendations were accepted by the association and referred to the Conseil supérieur de l'enseignement technique, which endorsed most of them. The minister of Industry and Commerce, M. Hérisson, then asked the Conseil supérieur to draw up a new law governing the schools, saying that "the Arts et Métiers can now be considered to be a kind of secondary teaching" situated between the higher primary schools and the Ecole centrale: "We will have an uninterrupted continuum of professional establishments at all levels."[21]

The resulting law of 4 April 1885 raised admission standards by requiring candidates to have mastered all of elementary geometry, algebraic equations to the first degree, notions of history and geography, and a better knowledge of French—in short, the programs of the new higher primary professional schools, which were to furnish students to the Arts et Métiers. These requirements saved the latter from having to provide about six months of remedial work during the first year of studies. The revised program included analytic geometry, industrial hygiene, some history and geography, more French, and earlier specialization in the workshops. The requirement of analytic geometry raised the level of mathematics sufficiently to give gadzarts a better chance of passing the entrance examinations for the Ecole centrale.[22] Thus the schools were securely established as secondary technical colleges capable of preparing their best students for higher engineering and scientific training. The main objection to the law was that the revised official goals of the schools—"to train skilled workers capable of becoming shop supervisors and industrialists versed in the practice of the mechanical arts"—still described the programs and professional achievements of graduates in working-class and artisanal terms. It seemed ridiculous to teach a curriculum that could take a student to the Ecole centrale and still refer to graduates as "workers."

Despite their accomplishments, many old boys clung to the vestiges of their popular origin and working-class backgrounds, resisting any upgrading that they considered an imitation of the Ecole polytechnique and other fancy schools. Arbel and Trotabas were the most prominent spokesmen for this position. In a speech to a class reunion at Aix, Arbel opposed the raising of entrance standards in the law of 1885, providing us with an insight into the mentality of the self-made man in France:

Messieurs, I have reached the end of my industrial life, in which events have placed me in a high position and from which I have been able to judge men and their role in society. I have come to be grateful to the Providence that permitted me, a young man deprived of fortune, to enter the Ecole d'Arts et Métiers where I received a modest but solid education. Every young man ought to receive such an education, one which opens the way to the highest positions in life. . . .
You will forgive me when I tell you that I am opposed to the thought expressed by very good minds: 'raise the level of the programs of our schools.' I answer: no. I wish to conserve the school of the weak, of the needy, of the child of the worker who dreams of making his son a foreman. I wish to conserve the school of the widow who has scrimped and saved in order to pay for the lessons which her son has taken in order to prepare for the departmental entrance examinations. I do not think that anyone will gainsay me when I state that a student who has passed his final examinations after three years of study has done very well indeed and ought not to be expected to do more.[23]

By the mid-1880s the Arbel position was fast becoming a lost cause. The final step in the association's transformation came with the presidency of Hippolyte Fontaine from 1885 to 1886 and 1889 to 1890. A pioneer in the field of central electrical generation and distribution and one of the most distinguished graduates, he had sided with the conservative, noninterventionist wing of the organization during the 1870s, but he now saw clearly the growing importance of science to industry, and he became a partisan of the upgrading of programs and of increased political activity on the part of the association. In a speech to the banquet at Saint-Etienne at the beginning of his presidency in 1885, Fontaine reaffirmed Poulot's policies and refuted the Arbel position of maintaining the status quo in the schools' programs: "Our comrades, whose good intentions I respect, do not take into account the progress being accomplished in industry and in all degrees of education in other countries. If our schools remain stationary, they will quickly fall behind and will fail to render the services that the State and the parents expect of them." Referring to "a sea" of newly created higher primary professional day schools capable of training skilled workers and foremen far more cheaply than the Ecoles d'arts et métiers, he warned that "the detractors of our schools will have a field-day in budgetary discussions and will succeed in reducing or even suppressing the credits voted for the Ecoles d'arts et métiers." He added that the schools would always be open to modest social groups because the advent of compulsory and free public primary education had made instruction more accessible to all. Fontaine concluded:

The Ecoles d'Arts et Métiers must train students who are increasingly well educated, who are abreast of the latest developments in applied mechanical sciences, and who are able to advance themselves through sustained effort to the highest ranks in French industry. Without that, the state would be wrong to spend 5000 francs per student.

My friends, I drink to the prosperity of the Ecoles d'Arts et Métiers and to the slow, methodical, rational but continual rise in entrance examinations, programs and teaching, both in their theoretical and practical aspects.[24]

Poulot and Fontaine had committed the association to work for the steady upgrading of the programs and credentials of the schools. This meant portraying the gadzarts as an engineer who had bridged the gap between the mechanical culture of the skilled worker and the abstract education of the polytechnicien, the ideal type of industrial man, combining theory and practice, thought and action.

Fontaine was also anxious to secure "the unity of all gadzarts" under the suzerainty of the alumni association. The executive did not like the provision in the law of 1885 that established a *brevet supérieur* above the ordinary *brevet* of the schools and that also denied the right to use the name "ancien élève des Ecoles d'Arts et Métiers" to anyone who had not completed his studies. The association stressed "the equality of all gadzarts" and made a point of accepting those expelled from the schools for resistance to the "Strass" (administration).[25] It also vigorously opposed the separation of the three divisions undertaken by the Ministry of Commerce in 1873 to prevent the hazing of freshmen by senior classmen. It considered the separation a violation of the solidarity of all gadzarts and of the duke's idea of mutual teaching: "The students are walled-in, cloistered; there is no camaraderie, no more emulation, no more mutual learning, and, consequently, no more esprit de corps."[26]

During the 1890s the association won a series of important reforms. Advancement of the Ecoles d'arts et métiers to the secondary level in 1885, combined with the victory of the Ministry of Commerce in winning sole authority over the Ecoles pratiques de commerce et d'industrie in 1892 and the four Ecoles nationales professionnelles in 1900, gave Commerce a complete school system and made it possible to further upgrade the Ecoles d'arts et métiers.[27] In 1897 the ministry ended the separation of the three divisions and created a board of governors, of which 15 of 23 members were graduates. The National Assembly voted 206,000 francs for the electrification of the schools and their workshops and for the introduction of a complete course in electricity, including laboratories.[28]

In the law of 11 October 1899 the minister of Industry and Commerce, Alexandre Millerand, raised the science and mathematics content of programs and liberalized and demilitarized the internal regime of the schools. To be admitted to the schools a candidate had to possess the diploma of the Ecoles nationales professionnelles, the Ecoles pratiques d'industrie, or the Ecoles primaires supérieures. A new school was opened at Lille in 1900, Cluny was converted into an Ecole d'arts et métiers in 1901, and planning began for a third at Paris (1912).[29] The culmination of a decade of reform came with the award of the much coveted engineering diploma in 1907 and the elimination of the word *worker* from the revised prospectus of 1909.

Despite general progress in the affairs of both the schools and the alumni association (a budget of 1 million francs and a national organization consisting of 106 regional branches),[30] there were some problems of unemployment among graduates around the turn of the century caused by the stagnation of the mechanical, metallurgical, and railway industries. Unemployment reached a peak among *sociétaires* in 1905 at 4.5 percent.[31] Moreover, the annual number of candidates to the schools rose very slowly from 1,000 in the 1880s to 1,300 in 1905, while the number of schools had grown from three to five (300 to 500 places annually). The opening of the Paris school in 1912 would only make matters worse.[32]

Doubling the number of Ecoles d'arts et métiers from 1900 to 1912 coincided with the creation of engineering schools in 15 university science faculties, several of which (Grenoble, Lyon, Nancy, and Toulouse) were growing rapidly. Moreover, the Jesuits opened the Institut catholique d'arts et métiers at Lille in 1898, and the Christian Brothers, the Ecole catholique d'arts et métiers at Reims in 1900.[33] This meant much increased competition for executive engineering posts in industry, the kind of posts that gadzarts were used to attaining through job promotion. Many younger graduates objected to the expansion of the Ecoles d'arts et métiers at the very moment when so many other schools were coming into existence.

The new school that opened at Lille in 1900 was the first to be established since 1843, despite repeated efforts on the part of Lille and the northern departments to obtain one. By the 1880s the Arts et Métiers were turning away 300 qualified applicants annually for lack of space in the three schools. In 1878 the industrialists of the Nord sent the minister of Commerce a petition signed by 2,495 industrialists, 10 industrial societies, and 148 municipal councils in the Nord region. The minister, M. Tirard, traveled to Lille to say yes: "Seventeen of our most important cities have sought to obtain an Ecole d'arts et métiers. . . . The need is clear: industry is demanding skilled workers, well-trained foremen and intelligent technicians. It is in the national interest to put an end to a situation in which we are turning down qualified young men for admission to these schools."[34]

Tirard's announcement, followed by the government's decision to establish the school, inaugurated a period of 20 years of wrangling and delay; the central government, Lille, and the Department of the Nord quarreled over financing, a quarrel that became more bitter as construction costs mounted. The architect had designed a great stone monument to science, an immense cloister with mighty columns trumpeting the glories of modern technology. When the school eventually opened in 1900, none of the shops was ready, and the kitchen and dining hall had been relegated to an airless basement in order to save money on construction costs. The students lived, worked, and slept in the odor of "perpetually boiling beef" that suffused the building. But the mighty columns remained, untouched by the budget cuts that had affected every other aspect of the school.[35]

Cluny officially became an Ecole d'arts et métiers in 1901 but in fact had been one since 1891. In that year the old Ecole normale supérieure de l'enseignement spécial had been closed and the facilities transferred from the Ministry of Public Instruction to the Ministry of Commerce, which converted it into an Ecole pratique d'ouvriers et de contremaîtres. It had virtually the same programs as the Ecoles d'arts et métiers and drew its students and faculty from the same sources. In 1901 the Ministry of Industry and Commerce transformed it into an Ecole d'arts et métiers and awarded the *brevet* of the schools retroactively to the graduates of the previous years.[36]

Millerand's plan to open a sixth school in Paris was opposed by many gadzarts. Regional groups, especially those of Bordeaux, Charleville, Dijon, Montbéliard, and Nancy, were openly hostile to new creations and criticized the association executive for not taking a stronger stand against the establishment of new schools, especially Paris.[37] The executive hesitated, however, to oppose the ministry's decision concerning Paris, for it saw the advantage of having a school located in the neighborhood of prestigious educational institutions and scientific laboratories. Moreover, it was working hard to obtain an engineering diploma and did not want to offend ministry officials; indeed the engineering diploma seemed the only solution to the increasing competition from other schools.

During the nineteenth century the absence of a diploma was compensated for by the lack of competing technical schools and the chance to get ahead on the job. But the appearance at the turn of the century of university engineering schools threatened to close off traditional avenues of advancement for the gadzarts. Once engineering was monopolized by the universities, as in the United States by the turn of this century, or by higher technical schools, as with the Technische Hochschulen in Germany, or by a combination of both as in France, the Ecoles d'arts et métiers were likely to end up as schools for technicians, stuck in the middle levels of the industrial education hierarchy and indistinguishable, for example, from the Ecoles nationales professionnelles. In 1902 Ernest Pantz, president of the association, predicted "eminent peril for the Ecoles d'arts et métiers" if traditional avenues of advancement were closed to gadzarts and to the schools.[38]

The rapidly growing association mobilized its 106 branches into a letter-writing campaign in quest of an engineering credential. Its representatives doubled their efforts in the Conseil supérieur de l'enseignement technique. Finally, on 22 October 1907 the minister of Industry and Commerce, Gaston Doumergue, signed into law the new diploma *ingénieur des arts et métiers*. Henceforth, after a comprehensive examination at the end of the three-year term, about 60 percent of graduates earned the diploma. The remainder had to be content with the old school *brevet* until, in 1920, all graduates received an engineering credential.[39]

In 1909 the goals and programs of the Arts et Métiers were revised in line with the new diploma. Traditionally the schools had trained skilled workers and foremen capable of becoming production managers and industrialists; now the pejorative word *worker* was dropped and the schools' purpose was "to train production managers and industrialists specializing in the practice of the mechanical arts."[40] The students studied differential and integral calculus, mechanical and electrical physics, metallurgical chemistry, and more French language and liberal arts.

As it happened, gadzarts' fears concerning unemployment proved to be exaggerated. The applied science institutes that most actively turned out engineers before 1914—Grenoble, Nancy, Toulouse, and, to a lesser extent, Lille and Lyon—graduated mainly chemical and electrical rather than mechanical engineers.[41] In the decade before the war the arms race greatly stimulated the mechanical and metallurgical industries and therefore the demand for mechanical engineers. Gadzarts had few problems finding jobs or obtaining promotion during these years.[42]

Thus it is difficult to know whether the engineering diploma had any immediate impact on the schools' status or the graduates' employment prospects. Despite the new credential, the Arts et Métiers were still associated with the secondary echelon of the socially inferior primary system, whereas all other branches of engineering studies were established in higher education. Most of the students entered the Ecoles d'arts et métiers at the age of 16 or 17 and graduated in three years at the age of 18 or 19, roughly the same age as those obtaining the *baccalauréat* after study in the lycées or collèges. Around 1900 about 15 percent of gadzarts continued their studies in such higher schools as the Ecole centrale or the Ecole supérieure d'électricité. Forty-three percent of the students at the Institut électrotechnique de Grenoble in 1910 were gadzarts. Parliament voted 50,000 francs in 1890 to provide scholarships for graduates of the Arts et Métiers to attend the Ecole centrale.[43]

The question of gadzarts' performance in the higher schools of engineering provides us with an insight into the strength of the programs of the Ecoles d'arts et métiers and the value of combining training in theory and practice in classroom and workshop (by the turn of the century seven hours in class and five in workshops). As early as the 1860s Arthur Morin observed that several gadzarts were being admitted annually to the Ecole centrale. Though on entry they were usually ranked lower than their competitors from the lycées in the pure sciences and mathematics, they almost always finished among the top graduates. Morin said that "their talent in mechanical drawing, their knowledge of form, and their ability to conceptualize a problem gives them a marked superiority over their competition in all questions having to do with the execution of a project." Once at the Ecole centrale, "most of the young men live in Paris in the most modest circumstances amidst fellow students who are richer and sometimes less assiduous than they, yet they have all upheld the honor of the school. Today, most of them are engineers who have reached very high positions in their fields."[44]

In 10 *promotions* at Centrale from 1892 to 1901, there were 87 gadzarts, of whom 77 obtained the school diploma and 5 the certificate of studies; only 5 did not finish their studies. In every category the gadzarts performed well above average. In the 4 *promotions* from 1898 to 1901 (300 per promotion), 31 graduated, of whom 12 (40 percent) finished in the top 10 of their class and 3 ranked number one. Eight gadzarts in the promotion of 1900 ranked as follows:[45]

Rank on Entry	Rank on Graduation
13	6
17	1
19	85
26	2
91	19
103	5
152	3
175	194

The Ecole supérieure d'électricité was founded privately in 1894 and very soon acquired an international reputation. It recruited from among engineers from the *grandes écoles* and the universities who wanted advanced specialization in electrical engineering. In the face of such high-level competition, it is surprising at first glance to find that graduates of the Arts et Métiers formed the largest single body of students at the school. Of 2,804 students from 1894 to 1925, 408 (one seventh) were gadzarts:[46]

Engineers from the Ecoles d'arts et métiers	408
State and military corps of engineers	407
Universities: licenciés ès sciences	361
Ecole centrale des arts et manufactures	196
Ecole polytechnique	194
Universities: engineers from the applied science institutes and other technical schools	114
Ecoles des mines, ponts et chaussées, and so on	32
Admitted through competitive examination	937
Foreigners	155
	2,804

In 1909 Emile Boutmy, president of the Board of Trustees of the Ecole supérieure d'électricité, noted that gadzarts were admitted to the school on the basis of competitive examinations alongside students from the Ecole polytechnique, the Ecole centrale, the science faculties, and the engineering institutes. "They bring with them a special style which we like very much. They are much more experienced in the shop, more competent and skilled in the execution of projects. This gives them a superiority over their comrades who have received a more advanced theoretical instruction but who attach far too little importance to practice."[47] The director of the school, Paul Janet, a *normalien* and a distinguished scientist, reported in 1912 that of 106 graduates that year, the first, second, and fourth ranked were all gadzarts. "Aside from their incontestable superiority in all things practical, one wonders how they are capable of grasping the frequently delicate theories of modern industrial electricity." He attributed this to the combination of theory and practice taught in the Arts et Métiers and the students' capacity for hard and sustained work.[48]

In 1921 Janet described the gadzarts at "Supélec" as "young, hard-working, in good physical condition, very methodical workers … [knowing] how things are made and the results to obtain."[49] They were strong in mechanics, physics, and industrial design but rather weaker in mathematics, in which they frequently had to take a year of preparatory work in the advanced sections of the lycées to qualify for the *grandes écoles*. Once having done so, they were among the best students in the school. In the class of 1920, 2 of the top 6 were gadzarts. In 1931 the director reported that of the 31 gadzarts who had just graduated, 5 had made the top 10, placing first, second, and third, and 17 had made the top 25.[50]

The excellent performance of gadzarts at such schools as the Ecole supérieure d'électricité in competition with graduates of the higher schools suggests that the "superiority" of the polytechniciens and perhaps the centraliens was more social than real. Though the latter were a year or so ahead in mathematics, they were two or three years older on the average, having reached Polytechnique or Centrale after two years of preparatory studies in the lycées at the age of around 19 or 20 and graduating two or three years later. In contrast gadzarts graduated at 18 or 19 years of age. Their high level of performance, despite their disadvantage in age, suggests that the combination of theory and practice taught at the schools constituted a valid pedagogical approach. The application of the science and mathematics learned in the classroom to the workshop programs provided for day-to-day observation of physical and mechanical phenomena, providing concrete images of the knowledge obtained in textbooks, calculations, and formulas. Hence the students managed to achieve a spatial comprehension—the ability to "see" objects represented in plans in their real dimensions—which provided the basis for mathematical understanding. With some additional course work, they could move into such specialized fields as electrical and aeronautical engineering.[51]

The study of mechanical drawing and machine design was of particular importance in this regard. Because of their training, gadzarts understood at once the operation and control of machines and the principles underlying their functioning. Hence they could design machines and/or adapt them to new uses. This design capacity is what made them engineers, not just skilled mechanics, long before they obtained the engineering credential. The gadzarts contrasted their knowledge of machine design with the book learning of the polytechniciens and centraliens, who spoke elegant French but who were considerably less skilled than they in the language of design, and they concluded that they, the gadzarts, were "the true engineers."[52]

However, the rise of large centralized corporations during the first decades of the twentieth century and the relative decline of small companies (in which gadzarts had done well) forced many to compete for managerial positions in large firms with the graduates of the big technical schools, the science faculties, and increasingly the Ecole supérieure de commerce and related business schools. Despite the upgrading of the programs and diplomas of the Ecoles d'arts et métiers in the reforms of 1885, 1899, 1907, and 1909, they were still modest provincial technical colleges whose graduates were specialists lacking in the language and business skills thought to be important in managerial circles.

During the early decades of this century, therefore, questions of status preoccupied the alumni association and led to tensions between younger graduates, often salaried employees living in the provinces, and the old boys, many of whom were self-made men residing in Paris, who expected their younger "comrades" to follow in their footsteps. Serious divisions within the association were avoided only because the executive, beginning with the presidency of Poulot in the early 1880s, introduced reforms partially decentralizing and democratizing the organization. In addition to its lobbying and public relations activities, the association played an increasingly active role in the governance of the schools and in student life, portraying itself as the embodiment of the "Esprit Gadzarts."

The campaign against *fonctionnarisme* helped to unite all groups within the association. During the years before the Great War, a new variant on the theme appeared with the introduction of Taylorism, a theory of scientific management proposed by Frederick W. Taylor, the American engineer. Although this theory generally had a limited impact in France, it appealed to many gadzarts and other industrial engineers because of its emphasis on industrial efficiency and the primacy of the engineer over the administrator in the production process. Taylor's goal was to eliminate all wastage of time and motion in the factory. This meant that the planning office, staffed by engineers and draftsmen, became the center of the workshop, minutely dividing labor into specialized tasks, controlling its every movement, and ensuring the standardization of machines, tools, and all plant operations. Some gadzarts concluded that the Taylor management system of "parallel hierarchies," as opposed to the traditional semimilitary command structure of the industrial corporation, would ensure greater productivity and give them increased control over the workers, thus enabling them to become effective arbiters between capital and labor. Such a system had an obvious appeal to gadzarts, who had always felt trapped between the workmen (and their unions) and the bosses.[53]

Taylorism and the campaign against *fonctionnarisme* combined with a tradition of loyalty to the corporation (the source of productivity and of social advancement for the engineer) to form part of the free enterprise mythology that held gadzarts together and brought them closer to engineers of other backgrounds. For many years the graduates of the Ecoles d'arts et métiers had formed about 20 percent of the membership of the Société des ingénieurs civils (organized in 1848), an organization composed mainly of centraliens. In 1883 Louis Martin (Châlons, 1837), a distinguished railway engineer, was the first gadzarts to be elected president. He was followed in 1902 by Jules Mesureur (Châlons, 1850), a retired manufacturer of plumbing supplies and water systems (though he died soon after and never served). Since then nine more graduates have been elected president of the society, a sign of the collaboration of centraliens and gadzarts in professional activities and industry over the years.[54]

By the end of the nineteenth century the alumni association's annual banquet and ball were attracting an increasing number of dignitaries. Félix Faure and Emile Loubet, presidents of the Republic, put in appearances to decorate successful industrialists and officers of the association with the Legion of Honor.[55] The ministers of the Army and Navy were as prominent as those of Commerce and Public Works, for the schools were an important source of technical officers, especially in the navy, where two thirds of the technical officers and technicians were graduates.[56] Indeed just about everyone paid more attention to the gadzarts as the arms race heated up. Speeches at the alumni banquets were fulsome in praise of their intelligence, toughness, and self-reliance. Graduates of the Arts et Métiers were very proud indeed of their achievements: the building of the big companies and the national transportation network, the inventions and new industrial processes, the engineering diploma, the flattery of the republican politicians, the big guns that were protecting the country against the German menace.

In all the speeches and writings of the graduates, only one, a speech by Léon Dufès, director of the Fives-Lille machine-construction plant near Saint-Etienne, spoke of the dangers of:

an armed peace, crushing taxes, . . . a colossal hardware out of all proportion to man's needs, . . . an enormous overproduction leading to ruinous, sometimes bloody conquests and of periods of intense economic crises which create frequent and regrettable conflicts between capital and labor.[57]

But nobody took up the cry from the remote industrial town of Saint-Etienne. For most gadzarts life in the Belle Epoque was at once too busy and too comfortable to worry very much about abstractions: "Le champagne pétille dans nos verres, les cigares s'allument, les têtes commencent à s'échauffer," said one graduate describing a Paris alumni banquet.[58] He might have added, not without a touch of pride, that the gadzarts had come a long way from the *boîte à fumée.*

7

The Ecoles d'Arts et Métiers Since 1914

Méfiez-vous du mot "statut" quand vous le verrez; vous pourrez parier à neuf chances sur dix que c'est un statut malthusien. Il faudrait peut-être faire un statut des status pour être sûr qu'on nén fasse pas d'autres.

Alfred Sauvy, *La Montée des Jeunes*

The Arts et Métiers had made substantial gains by the eve of World War I. Since 1899 programs had been upgraded, an engineering diploma obtained, and the number of schools doubled. The alumni association had grown from 2,000 in 1880 to almost 9,000 in 1913. The authority of the Ministry of Commerce over technical education had been greatly augmented by the acquisition of about 40 Ecoles pratiques d'industrie and 4 Ecoles nationales professionnelles. These provided the ministry with a complete hierarchy of technical and professional schools and were an excellent source of well-qualified students for the Ecoles d'arts et métiers. The best students of the Arts et Métiers could in turn continue their education in such higher schools as the Ecole centrale (under Commerce) and the Ecole supé-rieure d'électricité, neither of which required the *baccalauréat* for admission. As the arms race heated up, gadzarts played an important role in military preparations because of their close association with the mechanical, metallurgical, arms, and transport industries. This gave the alumni association an entrée into the Ministries of the Army, Navy, and Public Works as well as into the Ministry of Industry and Commerce.

Yet despite such gains, many politicians still thought of the Arts et Métiers as apprenticeship schools for skilled workers. When war came in 1914, 6,500 gadzarts were mobilized and sent indiscriminately to the front lines, where 1,100 were killed in the first year.[1] Centraliens and polytechniciens were given special dispensation as necessary technical personnel. When the generals and politicians were finally made to realize that they were slaughtering the young mechanical engineers who were indispensable to the fighting of a war of attrition, the gadzarts were recalled from the front. Henceforth they would play a major role in placing the economy on a war footing. As we will see in chapter 11, gadzarts participated in the mobilization of the railways for war, in the production of armaments, and in the rationalization of the manufacturing process in order to increase output. They made their most important contributions in the new field of aviation, in which they were responsible for the design, testing, and execution of virtually all of the airplanes produced during the war. They were the technical directors of the six most important companies manufacturing airplanes, airplane motors, and dirigibles. Because of their role in the war effort, they emerged from the struggle with enhanced prestige.

During the war the school at Châlons was badly damaged, Lille was occupied by the Germans and used as a hospital, while Aix, Angers, and Cluny stayed open part of the time but saw considerable deterioration of equipment and material. For reasons of economy, the government decided in 1916 to close the schools for the duration of the war, but thanks to Edouard Herriot's intervention in the Chamber of Deputies, the schools were reopened on a partial basis in 1917. However, only Paris survived the war in good shape.[2]

The Interwar Years

In 1920 Prime Minister Alexandre Millerand decided to transfer the Ecoles d'arts et métiers and other technical schools from the Ministry of Commerce to the newly created Ministry of National Education (formerly the Ministry of Public Instruction). As the former minister of Commerce from 1899 to 1901, Millerand had enjoyed an active and unusually long term of office by the standards of the Third Republic. He had presided over a major expansion of the technical schools under his ministry, including the doubling of the Ecoles d'arts et métiers (the planning for Paris began during his term). But in 1920 he was seeking to unify the schools scattered under various ministries under the Ministry of National Education. This led to strong protests from the alumni association and from the Association française pour le développement de l'enseignement technique, an organization founded in 1902 to act as an umbrella for the technical education groups.[3] Millerand then decided to create a semiautonomous technical education division within the ministry under an undersecretary of state and a permanent director. It was constituted as one of the four main divisions within the Ministry of Public Education: higher, secondary, primary, and technical-professional.

The first undersecretary was Gaston Vidal, an able man who enjoyed an unusually long tenure (1920–1924) by the standards of the Republic. Under his regime a number of changes were made in the Arts et Métiers. The age of entry was raised to between 16 and 19 years, the students graduating normally between the ages of 19 and 21. The teaching of mathematics was improved, and physical education, economics, and geography were introduced.[4] Although the program reforms of 1909 had introduced English, German, and "notions" of business administration and accounting, generally the gadzarts were deficient in business administration, which some felt hurt their chances for promotion on the job.[5]

In 1919 the government launched a program to modernize the shops and replace worn-out and outdated equipment. Students spent 5 hours of the 12-hour day in the workshops. These shops specialized in machine construction, assembly and installation, pattern making, metal turning, and foundrywork and were closely coordinated with courses in the applied sciences and mathematics. This meant the application of scientific knowledge to the manufacturing process and the scaling of production techniques in a way that was economically feasible and commercially exploitable. Students arrived at the Arts et Métiers from the Ecoles nationales professionnelles and the technical higher primary schools as skilled workers trained to the level of foremen; they had learned how to operate, maintain, and repair machines and had considerable knowledge of industrial drawing and design. In the Ecoles d'arts et métiers they trained to become production engineers (*ingénieurs de fabrication*), which meant that they learned how to conceive, design, and manufacture machines and machine tools from the selection and patterning of raw materials to the manufacture of the finished product. They learned how to construct and maintain industrial machines (the machines that make machines) and how to supervise the production process.[6]

The increased prestige of gadzarts resulting from their contributions to the war effort and improvements to programs and equipment during the postwar period inspired hope that the Arts et Métiers might be promoted to the university level. And yet by the 1930s the schools were apparently losing ground to the institutions of higher learning.[7] Gadzarts felt that their contribution to the war effort had not been appreciated and that their abilities were not being adequately utilized by either industry or the state.[8] Many soon came to regret the separation from the Commerce Ministry, for they were never very comfortable under the Ministry of National Education, despite the semiautonomous nature of the technical education division. The first years of the new administration went smoothly enough, but after Vidal's departure undersecretaries came and went in rapid succession for the next 16 years, which left great power in the hands of the permanent officials of the technical education administration. During their 88 years under the Ministry of Industry and Commerce, the technical education groups had been united in their dislike of the *universitaires*, and now these same mandarins had authority over the technical schools.

There were only two directors of the technical education division during the interwar period: Edmond Labbé (1920–1933) and Hippolyte Luc (1933–1944). Labbé, a former schoolteacher who had risen through the hierarchy of the primary system, had been director of the Ecole nationale professionnelle of Armentières in the early years of this century.[9] One of his first actions was to cut the number of state scholarships to the Arts et Métiers, replacing them with "honor loans," which made it more difficult for young men of modest means to attend. Of the 1,800 students in the six schools, only 800 had full scholarships, though some received aid from municipalities, departments, the alumni association, and other societies.[10] Before World War I, almost all students had received at least partial scholarships.

Labbé also decided that the rough young fellows of the technical schools needed more culture. He developed a program called "les humanités techniques," which he hoped would bridge the gap in prestige between classical culture and professional studies.[11] Though gadzarts had a side that yearned for the higher culture of the polytechnicien, they were mainly practical men with little feeling for the niceties of language and polished manners. Thus when Labbé spoke of molding "un gadzarts perfectionné," he rankled the average gadzarts "non-perfectionné" more than he could ever have known.[12] Gaston Vidal, who was well liked and respected, also managed to offend them by promising "to teach them how to speak French."[13] Labbé liked to talk about the schools' workshops in the quaint artisanal sense of training sturdy blacksmiths, mechanics, and turners. In speech after speech he evoked the image of a manly and paternal gadzarts placing a tool in the rough, uplifted hands of a worker, to the annoyance of the actual gadzarts who were involved in production engineering in the factory and planning office. Labbé was an endless source of clichés: his favorite was "Tout pour la profession, et par la profession," which implied a narrow professionalism in contradiction to his theme of "les humanités techniques."[14]

In 1924 Prime Minister Edouard Herriot, head of a new government of the left, failed to name an undersecretary of state for technical education, which appeared to demote technical education to a lower rank. Though he subsequently changed his mind and named an undersecretary, Herriot alarmed the alumni association by talking of transferring the Ecole d'arts et métiers at Cluny to his home city of Lyon.[15] Labbé envisioned the creation of a seventh school at Bordeaux. Soon everybody at the rue de Grenelle seemed to be talking about establishing new Ecoles d'arts et métiers, and the gadzarts, far from feeling flattered at their sudden popularity, felt threatened on all sides.[16]

The alumni association used its influence in the Conseil supérieur de l'enseignement technique to stem the tide for Bordeaux and Lyon, but when Labbé took up the cause of Strasbourg, a city just liberated from the Teutonic yoke and already possessing a school (an Ecole nationale technique) suitable for conversion, the association found itself fighting a losing battle. In 1925 the National Assembly voted to transform the school into an Ecole d'arts et métiers pending a special ministerial decree of implementation. The association argued for either the closing of Cluny as a quid pro quo or a general reduction in the number of annual admissions to each school from 100 to 80 pupils, in order to keep the total number at about 600.[17]

Of course the association could not have guessed that a regime would pass a law and then not execute it, yet the enabling decree never came. The Ministry of Public Education and Parliament squabbled over financing, and by 1933 the Depression had left so many gadzarts unemployed that all action was postponed indefinitely. The technical school at Strasbourg was converted after the Second World War into an engineering school ranking just below the Ecole d'arts et métiers in prestige.

The alumni association did have a legitimate concern about recruitment. During the 1920s the number of candidates to the Arts et Métiers remained as it had been in the decade before the war, about 1,300 per year.[18] This was adequate when there were only three schools, but now there were six of them. Moreover, 15 applied science institutes and other technical schools competed with them for students. The number of applicants began to fall in the late 1920s because of war losses and the low birthrate that ensued. The gradual absorption of the higher primary schools into the lycées and collèges threatened to divert a major source of students from the Ecole d'arts et métiers into classical studies. It would be disastrous, all said, to expand the number of Ecoles d'arts et métiers just when the existing ones were threatened with decline.[19]

The placement of graduates also proved to be a problem during the interwar years. Although they had little difficulty finding employment during the 1920s, graduates complained that big companies hired them as machinists and draftsmen at low salaries and only slowly promoted them to engineering positions. The younger graduates, "les jeunes promotions," organized themselves into a distinct group within the alumni association, demanding that the association improve its placement service and public relations efforts.

The young gadzarts' complaints had their echo generally within the engineering profession during the interwar years. It was widely believed that there were too many engineers in France and that the rise of the big multiunit corporation encouraged the growth of bureaucracy and discouraged private entrepreneurship. The word *cadres* came into usage in France during the 1930s to denote the emergence of a new class of salaried managers whose bourgeois status depended on diploma and position rather than on patrimony or private ownership.[20]

Such changes at first favored the advancement of engineers to executive positions, for they understood the relation of applied science to industrial performance. But as companies grew larger and more complex, the demand for commercial and administrative skills grew apace, as did the fortunes of business school graduates.[21] The overloaded programs of French engineering schools left little room for economics, business, or modern languages. The Ecoles d'arts et métiers, for example, burdened with science and engineering courses and extensive shop and laboratory work, offered only introductory courses in business and accounting and in English and German during the first half of this century. Graduates complained that they lacked the culture and the knowledge of management techniques that would enable them to compete more effectively for executive positions in industry.

In the early 1930s the alumni association reorganized its placement and public relations programs and actively cooperated with the various organizations affiliated with the Fédération des Associations, Sociétés et Syndicats Français d'Ingénieurs Diplômés (FASFID) in obtaining passage of the law of 10 July 1934. This law restricted the right to award the engineering diploma to 88 accredited engineering and technical schools as certified by the "commission des titres d'ingénieurs" created by the law. *Autodidactes* could still obtain the engineering diploma but only after considerable professional experience and a difficult examination taken at the Conservatoire des arts et métiers of Paris.[22]

With the coming of the Depression, employment prospects for engineers declined rapidly. By the autumn of 1931 the number of unemployed gadzarts jumped to 244, mostly young graduates, reaching a peak of 678 of 15,000 members (4.5 percent) by December 1934. It then began to fall as dramatically as it had risen. By the end of 1936, 372 were out of work, the lowest total since 1931, and by 1938 the figure had fallen to 187, or 1 percent. Of the 8,000 members under 35 years of age, only 20 were unemployed. By the end of 1941 French industry hired all but 92 members of the association, though between 800 and 1,000 members were imprisoned in Germany.[23]

The sudden rise of unemployment among graduates during the early 1930s caused the alumni association to lobby strongly for a drastic cut in admissions to the schools. In 1933 the undersecretary, Anatole de Monzie, agreed: "Il faut limiter le recrutement et valoriser les diplômes."[24] Labbé, who usually opposed the worst "Malthusianism" on the part of the association, was about to retire, and so the association got its way. The number of students admitted annually to the six schools was cut from 600 to 360. With the beginning of the arms race shortly thereafter, France badly needed mechanical engineers just when the Arts et Métiers, the principal supplier, had reduced their output by 40 percent. A subsequent effort by the directors of the alumni association to reverse this policy in view of the requirements of the arms race was successfully resisted by a coalition of regional groups and the younger graduates.[25] The irony was that France, with its millions of unemployed, had a shortage of skilled personnel in the years before the war. While the Ecoles d'arts et métiers sat half empty, France had to import skilled personnel.

For all the efforts of the alumni association "to validate the diploma of the schools," during the interwar years their programs tended to stagnate, and graduates found it harder to meet the mathematics requirements for entry into the higher schools without doing ever more preparatory work.[26] With the increasing nationalization of defense industries beginning in 1936, the state's bureaucratic methods were extended into industry. Everything concerned with national preparedness seemed to be organized into a corps having its own Ecole nationale supérieure in one speciality or another, and each school had its own *concours* for a limited number of places, its special points for certain schools, and its highly competitive examinations. This "diplomania" was an external manifestation of an educational process that was becoming more politicized than ever in France by the 1930s.

The alumni association spent much of its time after 1936 trying to clear away bureaucratic obstacles to the hiring of gadzarts in the various military technical corps. Only the navy, which had obtained most of its technical officers from the Arts et Métiers for over a century (four of whom held the rank of admiral by the interwar years), fully appreciated their value, and it was perhaps no coincidence that the navy was generally considered the most technically advanced sector of the French armed forces during the 1930s.[27] The air force also recruited gadzarts but on a less regular basis. The Ecole supérieure d'aéronautique, which trained aeronautical engineers, admitted the top graduates into its preparatory year without examination. Of the 12 *promotions* from 1923 to 1935, the top student in 4 was a gadzarts.[28] But the Corps des officiers-mécaniciens de l'air, which trained technical officers for the air force, made it almost impossible for them to gain entry because of low pay, strict seniority regulations, and its habit of promoting aircraft mechanics on the job to technical officerships.[29]

The Ecole de l'air, a training school for pilots, kept gadzarts out until 1936 by requiring the *baccalauréat*. The armaments corps (*travaux d'armement*) made no provision at all for the admission of gadzarts; indeed, as late as 1938 it appeared not even to know that the Ecoles d'arts et métiers existed.[30] Thanks to the intervention of the alumni association, most of these oversights were corrected in 1939 and 1940, but by then it was too late. Ironically, only two decades earlier the gadzarts had contributed substantially to the great technical and industrial achievement of France in fighting and winning a long war of attrition and in becoming the most advanced nation in the world in the production of aircraft.

Many gadzarts openly disdained a regime that seemed to favor the bureaucrat over the man of action and that extended the domain of the former—thanks to the nationalization of the defense industries beginning in 1936—right into the sacred preserve of industry itself.[31] A. de Monzie, minister of National Education, told the annual meeting of the alumni association in 1933: "I believe in the good will of associations, but, you see, we are facing a very great evil. I believe that there is a growing spirit of defiance in industry against the public powers, against parliament and the government."[32] He was right. In 1938 G. Lavoisier (Lille, 1928), writing in the review *Ingénieurs Arts et Métiers*, blamed the government and the labor unions for the "strange torpor" and the "paralysis" that afflicted the recently nationalized aviation industry. He cited: "the frightening egotism, the exclusiveness, and the jealousy of the workers" who had been given higher pay and shorter hours by the Popular Front and still did not hesitate to strike in the spring of 1938. Second, the state had failed to establish clear-cut lines of authority, so no one knew whether the government or management was making decisions. The result was low output at high cost, increasing red tape, and "disorder, disorganization and demoralization" in the entire industry: "Now that big industry, and especially nationalized industry, has been placed under the power of bureaucrats, they will seek out men with diplomas and book knowledge rather than men of action; industry will be run by tired-out senior administrators, rather than by young men able to invent, to execute, to command." Lavoisier thought that the following reforms were necessary to restore the vitality of French industry and therefore the country:[33]

1. Freer play for men of youth, character, and daring.

2. The "social education" of workers. "Joy in work."

3. A single union uniting capital and labor.

4. Measures to encourage the family and raise the birthrate.

5. Hard work, the public good, national well-being.

During the late 1930s gadzarts talked increasingly about the failure of politicians to listen to engineers and about the need for technical elites to play a more important role in national life. Some turned to Catholic corporatism as an alternative to bourgeois liberalism and to communism and fascism.[34] For a time a number of graduates supported Vichy with its slogan "Work, Family, Fatherland." The alumni association followed a pro-Vichy policy from 1940 to 1942 under the presidency of Louis Bacqueyrisse (Angers, 1893). Bacquey-risse called upon the "Grande Famille" of 16,500 gadzarts to serve "the National Effort" with discipline, to be "the apostles of the New Order," and in so doing become "true Frenchmen contributing to the hard Crusade for the Public Good." He proclaimed Marshal Pétain the national savior: "Follow him, help him, love him."[35] In December 1941 he received a communication from 39 engineers of the Arts et Métiers, prisoners of war in Germany, declaring their entire devotion to Pétain and their willingness "to enter the service of the National Revolution on their return to the Motherland either in the [Condor] Legion or in professional activities." Bacqueyrisse appealed to Pétain to win their release (who seems to have done so) and circulated their letter, along with Pétain's encouraging reply, among gadzarts in other German prisoner-of-war camps.[36]

It is difficult to know how many gadzarts shared the sentiments of Bacqueyrisse and the executive committee. Many were temporarily seduced by the prospects of power returned to the chiefs, of unions tamed, and of bureaucrats put in their place, until it became apparent that the Vichy court was even more Byzantine than the republican one and that the regime after 1942 was a tool of the Germans. Most gadzarts continued to work hard, and there were plenty of good jobs and opportunities for promotion throughout the war. France was Germany's number one supplier, and the companies for which gadzarts worked (such as Renault) did most of the supplying. Some were active in the resistance: Pierre Beucher (Châlons, 1913) was Colonel Kléber, a leader of the national federation of resistance groups. A total of 353 gadzarts were killed in the war (80 while deported) and 1,000 were prisoners of war.[37]

In summary the interwar years were frustrating ones for the Ecoles d'arts et métiers. During the early 1920s their diploma was given equivalence to the *baccalauréat* for admission into the university science faculties. This apparent promotion in status actually confirmed the Arts et Métiers as secondary technical colleges at the very time when engineering studies had become firmly established in higher education in France, Germany, England, and the United States. This meant that the engineering diploma of the Arts et Métiers was little more than an admissions certificate to the *concours* of higher engineering schools such as the Ecole centrale and the Ecole supérieure d'électricité, though not the Ecole polytechnique and its affiliates, Mines and Ponts et Chaussées, which required the *baccalauréat*.

Thus the alumni association of the Ecoles d'arts et métiers had little success during the interwar years in its efforts to obtain an improved "statute" promoting the schools to higher education. But it grew rapidly and became more influential among gadzarts. Before the Great War three fourths of graduates belonged to the association; since the war to the present day almost 90 percent are members. During the 1920s membership grew to 15,000 of an estimated 16,000 living graduates (88 percent); it possessed a large capital from gifts and legacies of 5 million francs, and it had an annual budget of 1.25 million francs. The association published a technical journal, the *Revue technique*, the *Bulletin mensuel*, and the *Annuaire*. It maintained a retirement home on the Côte d'Azur, a pension plan for the needy, and a scholarship fund of 600,000 francs for students and for graduates continuing into higher education. It also spent 100,000 francs annually for political purposes and public relations.[38] Based on a network of regional groups at home and abroad, the association was well placed to take advantage of the reformist spirit of the postwar years.

The Arts et Métiers Since 1945

In 1943 the technical education committee of the alumni association drew up a program of reforms for the Ecoles d'arts et métiers for the postwar years. The committee was chaired by a distinguished alumnus, Jean Fieux (Cluny, 1902), a pioneer in the field of gyroscopic research. It recommended: (1) a fourth year of studies and the advancement of the Arts et Métiers to university status, (2) eventual reduction in the number of the schools and their transfer to major industrial and university centers (much discussed, never applied), (3) raising the admission age to 18 or 19 years, (4) continuation of the compulsory *internat*, (5) restoration of government scholarships, and (6) improvement of the preparatory programs in the Ecoles nationales professionnelles and introduction of a national system of lycées techniques.[39]

Fieux testified in favor of this program before the Langevin-Wallon Commission. All agreed on the need to create a system of lycées techniques, introduce a *baccalauréat technique*, and advance the Ecoles d'arts et métiers into higher education.[40] In 1946 the lycées techniques came into existence, and the programs of the Ecoles nationales professionnelles were upgraded. Both introduced a preparatory year for the *concours* of the Ecoles d'arts et métiers.

The assurance of a steady supply of well-qualified *bacheliers techniques* preparing specially for the Arts et Métiers made it possible to promote the schools to the university level. In 1947 a fourth year was introduced and the age of admission raised to 18 or 19 years. The schools were given a single name, the Ecole nationale d'arts et métiers. The five provincial schools at Aix, Angers, Châlons, Cluny, and Lille became "regional centers" preparing for a fourth and final year in Paris that would unite all students and train them in their specializations. During this year they were concerned mainly with industrial research and problem solving. Thanks to the new four-year program, the Arts et Métiers were much better able to keep pace with rapid developments in physical mechanics, aerodynamics, electronics, metallurgical chemistry, and thermodynamics. New courses in business administration and economics were also introduced.

Once the flurry of postwar reforms had been completed, the education system returned to old routines. The difficult economic conditions of the postwar years accompanied an assumption of zero growth, stagnant demand for technical personnel, and a general pessimism about the future of a world caught up in the cold war. By the early 1950s the Ecole nationale d'arts et métiers was still graduating only 360 engineers per year, and their percentage in the engineering profession was declining.[41] In 1956 France was training only 90 engineers per 1 million inhabitants, as opposed to 114 in Italy, 155 in the United States, 236 in the Soviet Union, and 237 in Great Britain.[42]

In 1956 the Union of French Metallurgical, Mining and Mechanical Industries launched a major investigation into the manpower needs of these industries, which encompassed 500 companies, 1,300 plants and offices, and 614,000 employees. It concluded that the industry required double the number of engineers and technicians than were coming annually into the job market. Of 35,700 *ingénieurs diplômés* working in these industries, almost one fourth (8,550) were graduates of the Ecoles d'arts et métiers. The union predicted an annual demand of 728 engineers from these schools, more than double the 1955 output of 337.[43] The *Commissariat du Plan* made similar predictions. As France had to compete with the Germans and others in the European Common Market that was coming into being, such deficiencies in skilled personnel were taken very seriously. The technical education division began to plan for the establishment of three new Ecoles d'arts et métiers at Bordeaux, Le Havre, and Toulouse.[44]

The association was, as usual, divided on the issue of growth. Some believed that the limitation on enrollments of the 1930s had been "Malthusian" and wanted to do whatever was necessary to meet the requirements of industry for 1,500 to 1,800 graduates per year. Others pointed out that the schools had already agreed to begin graduating 460 per year (instead of 360) between 1956 and 1959 (up 28 percent). They noted that this corresponded to the growth in the Ecole polytechnique (265 students in 1955 to 300 in 1957) and that the latter had hardly grown at all over the past century. As one member of the executive concluded: "Why should we risk our standing when others safeguard theirs?"[45]

Opponents of expansion also pointed to continuing problems of recruitment. The lycées techniques and Ecoles nationales professionnelles were not producing enough *bacheliers techniques* to meet the combined needs of the Arts et Métiers, the Ecoles nationales d'ingénieurs, and various other engineering schools and institutes in France. In 1961 the number of *bacheliers* in France was as follows:[46]

26,000	bacheliers	philosophie
15,000	″	sciences expérimentales
17,000	″	mathématiques élémentaires
3,000	″	techniques

Caught between the charges of Malthusianism on the one hand and by the stubborn resistance to change of many *sociétaires* on the other, the alumni association executive characteristically sought a compromise. It got the number of proposed new schools reduced from three to one. A new campus was opened near Bordeaux in 1964 in a modern scientific and industrial complex at Talence alongside the university science faculty. This brought the number of schools to seven, with a total of 2,400 students (300 in each of the six regional centers, and 600 in the fourth year at Paris). Thus while industry continued to call for 1,500 to 1,800 graduates per year, the school was producing about 600. The gap was filled by the creation of other engineering schools, especially the Ecoles nationales d'ingénieurs, so that the percentage of gadzarts in a rapidly growing engineering corps continued to decline. In 1982 there were 20,000 gadzarts of 250,000 French engineers, or 8 percent, down from 25 percent of school–trained engineers in 1920.[47]

Because of continuing recruitment problems during the 1960s, education officials and the alumni association executive agreed that the schools should seek a broader selection of candidates by appealing to *bacheliers* in mathematics and science from the classical lycées. In 1956 an option B was introduced for this purpose, which proved fairly successful in appealing to candidates who had just failed to be admitted to Centrale, Polytechnique, and other elite higher schools and had "belatedly discovered a practical bent." The high salaries offered to engineers by industry doubtless encouraged some to develop a practical bent. Those who passed the entrance examinations for the

Arts et Métiers received the *Diplôme universitaire d'études scientifiques* (*DUES*), equivalent to two years of university studies. This gave them the option to complete the *licence ès sciences* in a university faculty or transfer to the Ecole nationale d'arts et métiers.[48]

Option B aroused strong criticism from traditionalists in the association who saw it as a "planche de salut" for unsuccessful "taupes," an escape hatch for the grinds who, having failed to be admitted to Polytechnique or Centrale, were willing to accept either the Arts et Métiers or the *DUES* as a consolation prize.[49] Nevertheless, the experiment was cautiously undertaken in 1956, when 40 candidates were accepted under this option, as opposed to 380 under the regular option A (lycées techniques, Ecoles nationales professionnelles).

Indeed, recruitment under both options improved substantially during the 1950s, from 1,350 candidates in 1952 to nearly 2,300 in 1959.[50] By 1964, 100 of 800 applicants were admitted via option B and 500 of 1,600 under option A. The main problem with the two options was that part of the first year at the Arts et Métiers had to be devoted to teaching the *lycéens* more technology and the *bacheliers techniques* more mathematics. In 1963 the programs and goals were further upgraded and the school renamed the Ecole nationale supérieure d'arts et métiers (ENSAM). ENSAM provided an engineering education that was at once "scientific, technical and practical." Teaching was oriented toward production and research.[51] In 1964 the Ministry of National Education transferred ENSAM, the Ecole centrale, and other higher technical schools from the technical education division to the division of higher education (*Direction des enseignements supérieurs*). The alumni association protested that ending the connection with the technical education division would cut ties with industry and isolate the schools in the higher-education bureaucracy.[52]

In response a study commission established in 1966 by the higher education division recommended that the Arts et Métiers be promoted to *grande école* status.[53] This meant shortening the program from four years to three while increasing the preparatory course in the technical lycées from one to two years in line with the other *grandes écoles*. This would eliminate the awkward distinction between options A and B and more clearly define the status of the school. But just as the reform was about to be decreed, the student uprisings of 1968 took place, centered especially in the universities. These *évenements* so preoccupied the higher-education administration for several years thereafter that the advancement of the Arts et Métiers to *grande école* status was shelved. Not until 1974, eight years after the original recommendation, did the ministry implement the reforms.

But the delays were not caused solely by bureaucratic inertia. Serious divisions within both the alumni association and the Conseil de perfectionnement (board of governors) also delayed implementation. The association had always been divided between traditionalists and those who wanted to raise the standing of the schools. Some feared that the Arts et Métiers would become a gentrified engineering school recruiting from the bourgeoisie and increasingly teaching mathematics and pure science at the cost of technical studies.[54] This quarrel added to other tensions between the younger and older *promotions*, between the provinces and Paris, and between the "bosses" and salaried engineers.

Such disagreements, though typical of alumni associations everywhere, are more important in France because of association representation on the schools' administrative boards and councils. Since the late nineteenth century, for example, the alumni association has been represented in the Conseil de perfectionnement of the Ecoles d'arts et métiers, which lays down policies on programs, discipline, and entrance standards for the schools. Under the law of 1963 the board members included various inspectors and officials from the ministry, school directors, a professor from each school, the president of the alumni association, and four alumni association representatives designated by the executive. The minister of National Education also named two "important individuals" proposed by the association, two students elected by the fourth-year class in Paris, and six engineers and manufacturers proposed by leading engineering and industrial associations.[55] Thus the alumni association executive controlled five seats on the Conseil de perfectionnement, directly influenced the naming of two more, and may have had some influence on the candidates proposed by the other organizations. It had as much representation on this board (seven) as did the combined faculties of the seven schools. Moreover, the president sat on the permanent standing committee of the Conseil de perfectionnement, which participated in the routine operation of the school. Finally, each campus (regional center) possessed its own Conseil d'administration, which included the alumni association president or his designate and two association members. The Paris executive selected the representatives to the boards, even of the local centers.

In 1967 the Conseil de perfectionnement was partially reformed—the educational officials of the ministry received slightly less representation, the faculty somewhat more. The alumni association obtained one additional seat but lost the right to designate two outside individuals. The Savary Law reduced the power of alumni associations in the boards and councils of the higher schools in France. The reform will have a considerable impact on the future relationship between the association and the school.[56]

During the eight years of waiting for the implementation of the 1966 reform, there was considerable debate among gadzarts over the future organization of the school. Some favored a single institution in Paris, and some wanted two campuses, one in Paris and one in southern France. Another group wanted four centers located in the vicinity of science faculties in or near big cities: Aix, Bordeaux, Lille, and Paris, which meant closing down the centers in small cities and towns, Angers, Châlons, and Cluny. But this would leave the industrial northeast without a school. And so opinion gradually returned to the idea of seven centers, but three groups then developed. One wanted each of the seven schools to give the new three-year course and Paris to have no special status. Another wanted the six provincial campuses to provide the first two years of the program, with Paris uniting all the students in the third year. A third group—the one that ultimately prevailed—wanted Paris to provide the third year for all students but also give the first two years of the program to young people recruited from the Department of the Seine. This meant about 1,400 students taking the first two years in the seven centers, and 700 taking the third year in Paris.[57]

By 1970 the higher-education division still had shown no sign of implementing the 1966 project. Then new President Georges Pompidou, who succeeded Charles de Gaulle in 1969, and his prime minister, Jacques Chaban-Delmas, proclaimed a policy of encouraging industrial development and technology. After four years of fruitless efforts within the higher-education division, the alumni association executive appealed directly to the prime minister, who, to everyone's surprise, replied quickly and favorably. He formed a three-man "Comité des Sages" to explore ways to overcome deadlock in the Conseil de perfectionnement, disagreements within the alumni association, and the inertia of the higher-education division of the Ministry of Education. The committee recommended a plan quite similar to the 1966 proposal. The minister of National Education then put pressure on the higher-education division, and in April 1974 the Ecole nationale supérieure d'arts et métiers became a *grande école*.[58]

The decree of 12 September 1974 promoting the Arts et Métiers to higher status did not mean that the school advanced to the level of the four traditional *grandes écoles* (Polytechnique, Mines, Ponts et Chaussées, and Centrale), whose exalted position is well protected by tradition and by the influence of their alumni associations. Nor did it attain the status of such schools as the Ecole supérieure de physique et de chimie, the Ecole supérieure d'électricité, and the Ecole supérieure d'aéronautique, all located in Paris. These schools are among the best in France: they teach advanced scientific theory and mathematics along with applied science and engineering.[59]

The Ecoles d'arts et métiers belong in the much larger and more heterogeneous group of *Ecoles supérieures* in engineering and applied science, including the science faculties in the universities offering engineering programs. These schools and faculties, often referred to as "petites grandes écoles," are oriented toward the training of production engineers rather than research and administration. They lack the glamour of institutions training in high-level mathematics and pure science. However, ENSAM's programs are becoming increasingly scientific in nature and more research oriented, which may enable it to advance to a level roughly equivalent to the Ecole supérieure d'électricité and others. ENSAM is now ranked at the top of its category, followed by about 50 Ecoles nationales supérieures d'ingénieurs and related schools and about 20 Instituts nationaux des sciences appliquées, introduced in 1957. The 60 science faculties have generally less prestige than ENSAM and the Ecoles d'ingénieurs.[60]

To be admitted to ENSAM, the candidate must do two years of preparatory work beyond the *baccalauréat* in one of 42 lycées techniques offering the program "mathématiques supérieures et spéciales technologiques (T)." He or she studies mathematics, physical sciences, descriptive geometry, mechanics, French, and modern languages. Eighteen hundred candidates, 70 percent of whom possess the *baccalauréat E* and 30 percent of whom possess the *baccalauréat C*, then present themselves for 700 places open in ENSAM each year. An additional 60 places are reserved for holders of the *baccalauréat de techniciens*, the *brevet de technicien supérieur*, the *diplôme* of the two-year Instituts universitaires de technologie, and for self-taught individuals who have acquired a good knowledge of applied science and mathematics. Students coming through these "passerelles" do not possess the *baccalauréat* and are likely to be of modest origin.[61]

ENSAM is now a single school run by a director in Paris, assisted by a director of scientific studies and by various higher boards and councils. The six regional centers have identical programs, administrations, and faculty structures, and each has its own conseil d'administration. Since the reform of 1974, the professors have been assimilated to those of the technical *grandes écoles.* They are formally trained academics possessing the *certificat d'aptitude professionnelle de l'enseignement technique* (*CAPET*), the *agrégation*, and the doctorate; hence they are less likely than in the past to come from the world of business and industry.[62]

The program of the school covers a general range of the industrial sciences and engineering. In an age when materials must respond to space-age stresses and where machines communicate with each other, the gadzarts must have considerable scientific as well as engineering training. The program base is still mechanics (solids, fluids, vibrations, the technology of machines), as well as physics, chemistry, and mathematics, but new computer, electronic, and information courses are popular. Programs are generally open and flexible, emphasizing adaptation to a changing technology. The student spends 36 hours per week in classroom, laboratory, and shop, 32 weeks per year.[63] The third year in Paris is devoted to industrial research, mainly in developing new products for private industry, and to individual specialization. Paris has outstanding equipment, shops, and labs in aeronautics, aerospace, computer technology, electronics, and machine construction. Considerable emphasis is placed in all the schools on industrial planning and problem solving and increasingly on industial research. In 1982 the school had research contracts worth 8.7 million francs, mainly from industry.[64]

The gadzarts has always prided himself on his versatility, on combining thought and action, theory and practice. Now he sees himself as "un mécanicien-physicien, un concepteur-constructeur." He is more likely than in the past to experiment, do research, work in teams, speak English, and write reasonably clearly. In short he sees himself as a problem-solver, challenged by complicated problems in industrial production, dealing with vast amounts of information in the computer age, while working constantly with people and making decisions that have important social consequences.

Despite ENSAM's status as a *grande école,* the students still tend to be of modest origin and come from industrial backgrounds. As we will see in a detailed study in chapter 10, the sons of workers, artisans, and small manufacturers have been partially replaced by those of industrial employees and technicians and even those of senior executives and professionals.[65] In terms of careers, graduates have long since avoided the railroads and more recently the metallurgical industries. They are now beginning to desert the machine-construction companies for more glamorous high-tech industries (aerospace, computers, electronics, and robotics) or for teaching and research. One can do considerable work toward a doctorate in engineering in ENSAM.[66]

During the late 1970s the alumni association began to organize the celebration of the bicentennial of the schools for May 1980. It acquired the Duke de La Rochefoucauld's old estate at Liancourt and began to restore the dilapidated buildings and model farm (the château had long since disappeared). Thus the celebration at Liancourt on 18 May 1980 took place on the association's own property. In 1979 the organizing committee outlined the goals of the bicentennial:

Faire la démonstration éclatante du 'poids' des gadz'Arts dans le monde des Ingénieurs, en valorisant par la même leur image de marque étroitement associée à la formation technique originale et à l'esprit spécifique de leur Ecole.

Transmettre au générations futures un témoignage exceptionnel de fidélité à notre glorieux passé et à nos solides traditions, et de foi dans le rayonnement actuel et futur de notre Ecole.[67]

The French government proclaimed May 18 a day of national importance, issuing a commemorative stamp on the bicentenary. The association organized a series of lectures, made a slick public relations film on the schools, ghosted various articles in popular and professional journals, and arranged an official history of the schools, the *Livre d'Or, Bicentenaire Gadz'Arts*. The big day itself, attended by several thousand gadzarts, their families, and various dignitaries, saw a magnificent spread of French cuisine in a circus tent rented for the occasion, as well as various displays and demonstrations describing the life of the schools, their traditions, the activities of the association, and the achievements of graduates in business and industry.[68] The thirtieth "Gadzarts Games" were held to enliven the festivities. And of course there were many speeches and toasts to the memory of the duke and the glorious traditions. If 1880 saw the apotheosis of the Duke de La Rochefoucauld amidst the monuments the association had erected at Liancourt to his memory, 1980 saw the apotheosis of the alumni association amidst the monuments it had erected to itself.

Despite some tensions within the organization and a few splinter groups that appeared in the 1960s, the association is reasonably unified, well run, and very rich, especially in terms of its capital endowment. Centered in a luxurious mansion on the avenue d'Iéna in the sixteenth arrondissement in Paris (obtained in 1925), the Société des Ingénieurs Arts et Métiers is divided into 8 regional federations, over 100 hundred local groups within France, 3 in the territories, and 14 abroad. However, since the bicentennial celebration the Société des Ingénieurs Arts et Métiers and ENSAM have had to face a number of problems. A new socialist government, with Education Minister Alain Savary, proclaimed its intention to replace the Faure Law of 1968 with a more democratic law on higher education. This came just as ENSAM was beginning to enjoy its new privileges as a *grande école*. Indeed the alumni association's success in publicizing the advancement of the schools during the bicentennial celebration had mixed results. For the first time since Corbon and Le Play a century earlier, the schools and their association were criticized for having abandoned the original purpose as institutions providing opportunities for young people of modest origin.[69]

The Savary Law attempted to democratize the admission of students to higher education by allowing virtually free access to the first year of university. But during the second year (of three) selection was to become rigorous. This was combined with a general orientation of the universities toward business and industry through the establishment of "professional channels" (filières professionnelles), the precise definition of which await implementation by decree. Savary's plan to unify higher education under the Ministry of Public Education did not threaten ENSAM, which was already under the education ministry, but school officials were very unhappy with the proposal to do away with the *cours préparatoires*. In an effort to protect ENSAM's "prépas" in 42 lycées techniques, the Société des Ingénieurs Arts et Métiers cooperated wholeheartedly with the various organizations affiliated with the *Comité national pour le développement des grandes écoles* and the *Conseil national des Ingénieurs français* in a struggle against the Savary Law. These two federations, representing 200 *grandes écoles*, 250,000 engineers, and 50,000 higher technicians, constitute a powerful force determined to protect the privileges of the *grandes écoles*.[70]

It is ironic to see the alumni association, once the spokesman for the schools of the working man, allied with the Ecole polytechnique and other institutions representing the privileged sector of French education. The association contends that selection, if done fairly, is preferable from the working-class point of view to the random admission procedures of the universities, which admit *bacheliers* more or less indiscriminately. Though fewer workers' children gain admission to the *grandes écoles,* those who do are much more likely to earn a credential. In any case the Savary proposal, it argued, does not solve the problem of "la sélection par l'échec" but simply postpones it until the beginning of the second year of university. Such an extension of the selection process only works against those with limited funds.

The Société des Ingénieurs Arts et Métiers argues that it is in the national interest to train the best technical personnel possible. The *grandes écoles* produce superior engineers because they select better students, educate them more rigorously, and have closer contacts with business and industry through the school councils and alumni associations. "Professionalization demands selection and therefore selection is a beneficial law," the executive likes to say, with no suggestion of irony. For the association the Savary Law is "a leveling from below": "The best way to raise the level of scientific and professional culture in France is to bring forth the best people, who—and we repeat this intentionally—will be capable at every level of standing at the forefront of the cultural, professional and scientific progress of the nation for the benefit of all."[71]

The range of interest groups associated with the *grandes écoles* renders genuine reform of higher education in France extremely difficult. The law that finally passed in the autumn of 1983 poses no real threat to the schools located under various ministries or to the "prépas." It limits the representation on governing boards and councils of representatives of alumni, professional, and industrial organizations and does not give alumni associations automatic membership on such committees; hence such groups may lose part of their influence in the governing of the schools.[72]

There are other signs that the alumni association may be losing steam. Though membership reached 23,000 a few years ago, an impressive 88 percent, since then it has ceased to grow and may be declining. Receipts have leveled off while expenses have risen. The school's very success since World War II in improving its status and upgrading programs and diplomas, combined with the decline of old traditions, has rendered the association less indispensable. The end of the compulsory *internat* in 1964, the opening of a modern new school at Bordeaux during the same year, and the reduction of programs from four to three years in 1974 create environments unfavorable to the development of school spirit. While the association continues to encourage the old rituals and ceremonies, these are becoming increasingly hollow. Indeed the whole idea of the association as the protector of the "trads" seems irrelevant to many, especially among the younger graduates. The change of the name from the Société des anciens élèves to the Société des Ingénieurs Arts et Métiers underlines the point that the association is now, more than ever, a professional organization seeking to advance the interests of ENSAM as a *grande école* both in cooperation and in competition with the other higher schools.

A Culture of Refusal: The Boarding School and the "Esprit Gadzarts" During the Nineteenth Century

Ainsi que je voudrai, je pétrirai cette âme.
Comme un morceau de cire entre mes mains elle est.
Et je lui puis donner la forme qui me plait.

Arnolphe, speaking of Agnès,
Molière, *l'Ecole des Femmes,*
Act III, scene 3.

The Ecoles d'arts et métiers were probably the only European technical schools in which young workers were trained in boarding schools at public expense rather than in day schools or on the job. Because of the scarcity of middle-level technical managers during the early stages of industrialization, the schools recruited their students from the families of skilled workers and artisans. These were the "dangerous classes" to the bourgeoisie. Hence it is not surprising that the school administrations sought to destroy the students' working-class culture and acculturate them into bourgeois styles of life. Their purpose was to train technically competent and socially reliable middle-level managers for industry who would faithfully carry out the orders of the front office while instructing and supervising the workers on the factory floor.

The internal regime of the schools was a mixture of barracks and monastery, a harsh system against which the students constantly rebelled during the nineteenth century. The students, aged 15 to 18, developed an elaborate culture of opposition to the school authorities, including clandestine government and police, secret rituals and language, and a strict code of behavior that they passed from one generation to the next. The student culture was linked in many ways to their own popular and working-class origins and to their dislike of the narrow mold into which they were being forced by the boarding school regime. The alumni association, founded in 1846, prided itself on being the visible manifestation of the "Gadzarts Spirit." As it evolved into an educational lobby under the Third Republic, the association attempted to transform the spirit of resistance into a more positive image of industrial cooperation and teamwork. The evolution of the "Esprit Gadzarts" during the nineteenth century is the subject of this and the following chapter.

The Boarding School Regime

The Ecoles d'arts et métiers were boarding schools from 1803 to the end of the compulsory *internat* in 1964. The *internat* was a custodial institution embodying the principle of *clôture*, the confinement of students within school walls and their separation from the outside world. The *internat* was first introduced by the medieval Church in its seminaries for the training of the clergy; it was transformed and secularized under the Renaissance for the sons of aristocrats, and it was subsequently employed by reforming monarchs who saw it as an efficient means of preparing dutiful and competent higher civil servants. Napoleon adopted the boarding school format in 1802 in his new system of state lycées. The system of *casernement* and *clôture* suited his military temperament and met his need for highly trained and obedient civil servants and military officers.

As Emile Durkheim noted, the modern *internat* coincided with the rise in the West of an "esprit réglementaire" since the sixteenth century, which accompanied a "deep-seated horror of anything suggesting spontaneity, caprice and fantasy" and "the passion to impose a uniform and authoritarian system excluding all irregularities and expelling all rebels."[1] In *Surveiller et Punir* (1975), Michel Foucault argued that under the influence of the Enlightenment, a new kind of professional expert arose who believed that foolproof techniques of social control could be developed. In the school, the hospital, the factory, and the prison the "principe de clôture" helped ensure a closed and minutely regulated system of discipline and surveillance, of rewards and punishments, and of ranks and sanctions in order to fashion an acceptable social product.[2]

Napoleon, La Rochefoucauld, and the Internat

Though one can readily understand why Napoleon favored the *internat* for the new lycées and the Ecole polytechnique, it does seem surprising that he introduced such an expensive arrangement for the training of skilled workers and foremen in the Ecoles d'arts et métiers. A system of day schools or of learning on the job would have been far cheaper and would have reached more young people. In fact the development of the *internat* in the Arts et Métiers was something of an accident: the military institution at Compiègne already existed for orphaned sons of soldiers and veterans, so it made sense to teach the wards of the state something useful so that they could later support themselves and be of service to the state.

Had the *internat* not existed, however, Napoleon, Chaptal, and La Rochefoucauld might have introduced it anyway, for they recognized that in the future the state would need trained intermediaries, "sous-officiers pour l'armée industrielle," to fill the gap between the mass of unskilled factory hands and the bosses. La Rochefoucauld had already experimented with an apprenticeship school on his estate at Liancourt during the 1780s, and he realized that in the *internat* he could create a closed environment in which a new industrial man could be formed. The increasing presence in factory and shop of the educated worker would make possible "the simplification of processes, enhance productivity and lead eventually to a society of material abundance and harmony." Such men would build the greatness of France in the industrial civilization of the future.[3]

By the standards of the time there was an element of originality and even boldness in the duke's ideas. In an age when the notion of the industrial sciences and even of popular education was in its infancy, there was little precedent for the idea of public technical schools preparing skilled workers and technicians for industry. Indeed there wasn't much heavy industry in France at the beginning of the nineteenth century. The mechanical arts were associated with the vulgar working classes and were not therefore within the province of "honorable people." Hence they were bound to disdain any teaching method combining theory and practice, classroom and workshop. Moreover, artisans and workers were accustomed to apprenticeship and to learning on the job, and they were suspicious of school-trained workers and foremen. Hence artisans and craftsmen were reluctant to send their sons to the school.[4]

Thus most of the students at Châlons were the sons of soldiers or war orphans with little aptitude for technical studies. Many were scarcely literate. The school was plagued by thievery, rioting, rebellion, and evasion. Students were not afraid of being expelled and were not even intimidated by the school prison with its solitary confinement on bread and water. Because three fourths of the pupils either had no parents or had lost touch with them, the school could not enlist parental support or expel miscreants to nonexistent homes. The students, most of whom had no resources at all, were truly the wards and apprentices of the state.[5]

Jean-Jacques Labâte, the principal at Châlons, attributed the students' bad behavior to their background: "They come to the school from the lowest classes, having already acquired rude habits and vile sentiments and the most revoltingly vulgar manners."[6] He estimated that the average student arrived at the school with only 20 francs worth of possessions. One arrived with "only the clothes on his back" and another with "two dirty shirts in his knapsack." Few could afford the trousseau fee of 240 francs, and the government had to allocate 4,200 francs to outfit and equip the most ragged.

Despite such efforts, the disciplinary situation continued to deteriorate. In 1807 Labâte and La Rochefoucauld, backed by the Ministry of the Interior, decided to place the school under the military code. They organized pupils into regiments and companies under the command of selected older boys, ranked as sergeants and corporals. Serious offenses (*fautes graves*) were defined in military parlance as desertion, insubordination, theft, and mutiny. Punishments were, in the duke's words, to be "swift and exemplary" and in serious cases to lead to "one or two months in the county jail."[7] "Deserters" were hunted down by the police: "Any constable catching a deserter will receive a bounty of three francs to be paid on the return of the student to the school."[8]

The more serious punishments were reserved for offenses involving moral turpitude because of "the principle of corruption" involved.[9] Students frequently escaped over the crumbling walls of the old monastery at night to drink, carouse in cafes, fight with the townies (*la rixe*), and frequent the women of ill fame who, according to the duke, abounded in the slum neighborhood adjoining the school.[10] Such offenses led to automatic expulsion, but because this punishment incited little fear in the students and because the duke wanted them to regard moral turpitude with "the utmost horror," he argued that punishment should be "rapid, ignominious, severe, and exemplary." Under the military code of 1809, "Any student accused of serious breach of the regulations will be conducted to the city prison where he will be dealt with under the full severity of the law." This meant a jail term.[11]

To deter such offenses, the duke wanted "to isolate the school . . . and redouble surveillance." He successfully recommended raising the walls, buying and demolishing the adjoining houses, and instituting more police patrols. He increased the number of school proctors from four to six, replacing the *maîtres d'études*, usually ex-students, with guards (*surveillants*), mostly former soldiers or gendarmes. As the duke said, "The students have more respect for military men than for student proctors who are almost always as young as they are."[12] Finally, he established a military tribunal with power to "court-martial" students accused of major offenses. It was composed of the principal, assistant principal, two professors, two shop masters, the head proctor, two student sergeants, two corporals, and two ordinary students. In matters concerning morals, tribunal deliberations were strictly secret "in order to avoid public scandal," and the accused had no right of counsel or appeal.[13]

The walling in of Châlons in 1809 and the introduction of military proctors introduced a regime of incessant surveillance that deprived the students of privacy and of all outlets for youthful energies. The weakening of faculty influence by an all-powerful administration turned professors into bored and underpaid time-servers who took little interest in the life of the schools. These were essentially La Rochefoucauld's reforms, the product of the classical liberal ideas of behavior modification that influenced him.

In defense of the duke, one must point out that he apparently introduced authoritarian measures reluctantly. Originally the students were allowed to go into town fairly freely; they were supervised by assistants rather than by soldiers; professors were allowed the freedom to teach; and the internal regime was not excessively strict. La Rochefoucauld repeatedly tried to persuade the students to improve their behavior. He regularly visited the schools, counseling, aiding, and sometimes opening his purse to students in need. He also used his contacts to find jobs for the better graduates. He believed in appealing to the "honor" of each student and liked to make speeches in awards ceremonies in order, as he said, "to provide lessons drawn from events, admonish the students in areas needing improvement, encourage good dispositions, awaken enthusiasm, condemn inertia, and finally, to give to the awards a usefulness going beyond the distribution itself."[14] But this emphasis on personal example, on "emulation," as the bourgeoisie called it, failed to improve student conduct. Because of the ragged and untutored nature of the students, the lack of professors trained in the industrial sciences, the capricious attitude of the Emperor, and continuing disorder in the school, the duke felt compelled to introduce authoritarian measures. Despite this he was apparently well liked by the students. In an age when the upper classes looked down on working-class people and manual work, he at least took the students seriously and respected what they were doing.

The Restoration

The unruliness of students under the Empire was spontaneous and unpredictable rather than concerted and chronic. The long war between students and administration really dates to the Restoration and specifically to the period 1823–1830. In 1823 the minister of the Interior forced La Rochefoucauld to retire as inspector general of the schools and replaced Labâte with the Vicomte de Boisset-Glassac, an ex-army officer and a stiff-necked and narrowly religious man who had been a prefect of *La Congrégation* in 1821. One of his first acts as director was to refuse admission to a young Jew named Brissac, the son of a textile manufacturer from Lunéville.[15] He soon alienated everyone by insisting on regular religious observance and frequent communion. He inherited from Labâte a prefectural system in which the older and better students were chosen to act as "corporals and sergeants" to assist in maintaining school discipline. Boisset chose them on the basis of conformity to his own religious views rather than for intellectual ability and manual skill. This created an atmosphere of bitter hostility toward the administration. The students formed a secret society modeled on the republican *carbonari* of the time: there were passwords, clandestine meetings, secret messages, and nocturnal forays. At Angers the senior students formed an elite group of 10 called "la bande noire," who, wearing masks and armed with grappling hooks, ropes, and passkeys forged in metal shops, regularly escaped over the school walls during the dead of night. They broke into administrative offices, changing grades and placing a one before a zero for conduct, warning those who were theatened with punishment or expulsion. When discovered, they overpowered the proctors and left them bound and gagged, to be discovered the following morning. They enforced a rule of absolute silence. Stool pigeons (*mouchards*) were severely punished; even those falsely accused were not allowed to break the law of silence. The few who did so were ostracized by their fellows:

De noirs fantômes se glissent dans l'ombre
Sans aucun bruit, amortissant leurs pas.
Où vont-ils donc? dites-vous avec crainte.
C'est le secret de l'école des Arts.

Vous qui voyez rôder dans cette enceinte,
Ne dites rien, car se sont des gadz'arts.[16]

As tension mounted at Châlons in the years after 1824, students
began to forge *loupins*, "terrible iron weapons," which they stashed
away in hidden arsenals. The proctors, in constant fear of attack,
armed themselves. Boisset reported to the minister of the Interior in
September 1825 that "we have been on the alert for three days. . . .
The dangerous time comes at night during the quarter of an hour
separating the evening study period from retirement to the dormitories. It is during this interval that the troublemakers profit from
the crowd gathering in the corridors and in badly-lighted stairways
to perpetrate their misdeeds."[17]

The firing of a well-liked staff member led on 1 April 1826 to a
full-scale insurrection. According to a prearranged plan, 20 of the
toughest "anciens" attacked the proctors' quarters, hurling rocks and
iron objects at them, seriously injuring several. The students then
occupied the school. But they had put aside no food and were soon
besieged by troops sent by the prefect. Once the revolt had been
crushed, 31 students were expelled and 8 of the presumed ringleaders
were arrested and brought before the courts, accused of armed rebellion and malicious destruction of property. In rainy weather the 8
were marched on foot by the mounted police to Reims, 140 kilometers
away, handcuffed and chained to each other 2 by 2, sleeping in hovels
and eating prison food.[18]

The Duke de La Rochefoucauld quietly got them a lawyer and
provided them with spending money for the long march and the
months in prison in Reims. So harsh was the treatment accorded the
young men, still in their teens, that public sympathy shifted in their
favor. When the trial finally took place nine months later at the end
of December 1826, the eight were unanimously acquitted by the
court.[19] The duke was 80 years old. He hated the faction associated
with the *Congrégation*, and this was his last victory over them.

In the purges following the uprising of 1 April 1826, about half of
the senior class was expelled. For the remainder of 1826 the school
was quiet. Yet by February 1827 Boisset was reporting to the minister
of the Interior that he had been obliged to evacuate 18 proctors and
students from the school under military guard because "they are in
absolute terror of retribution from the rebels."[20]

In this tense situation the Duke de La Rochefoucauld-Liancourt
died (27 March 1827). He had become a hero, patron, and father

figure to the students, some of whom owed their education to his financial aid. Eight students were named by the family as pallbearers in the funeral procession. Once the procession had begun, however, the gadzarts were confronted by the police, who claimed that they must have official permission to act as pallbearers. The commanding officer ordered them to surrender the coffin. The gadzarts refused, backed by the dukc's son. Claiming that he was acting on orders from above, the officer then ordered the police to forcibly remove the coffin from the students. In the scuffle that ensued the coffin was jostled and fell to the ground, and the duke's body rolled out into the street in front of the entire procession. This sacrilege was widely publicized in the press and led to serious criticism of the government.[21]

Soon after, a memorial service for the duke was held in Châlons-sur-Marne. Students of the school requested permisson to attend. Denied this, a group escaped from the school and attempted to enter anyway. They were set upon by the police and a riot ensued. The students retreated to the school to lead a full-scale insurrection that was put down only when the army intervened. Once again the school was decimated by expulsions and resignations: the student body apparently fell from 450 to about 190 for a time, and the schools vegetated until the reform of 1832.[22]

The July Monarchy

The internal life of the schools under the July Monarchy was relatively quiet, thanks mainly to the reform of 1832 and the introduction of capable directors at Angers and Châlons. Under the leadership of Dauban, Angers experienced no major troubles until the Revolution of 1848 (which nevertheless cost him his job). At Châlons, Vincent and his successor, Le Brun, managed to avoid major difficulties, with one exception. On 20 January 1841 an entire division was punished because several of their number had insulted a proctor. At 7:30 the next morning, "With a screech of a whistle and a shout of 'down with the scum' . . . they fell upon their proctors and beat them with their fists. . . . They then broke into the school stores and into the office of the bursar and flung all his papers, registers, accounts, and some mathematics instruments that had been stored there into the corridor." They also stole 900 francs, of which only 175 were ever recovered. It took two brigades of gendarmes and a batallion of troops to restore order on January 21. When the troops withdrew, 50 students reoccupied the stores, and this time 200 soldiers were brought in to expel them. Damage from the two-day riot amounted to 6,000 francs plus the stolen money. The desperate director proposed to the Ministry of Commerce that parents of all incoming students henceforth be required to sign a contract stating that if their son was expelled from school for indiscipline, he would be obliged to serve three years active duty in the navy.[23]

Vincent reported in 1842 that of 908 students at Châlons from 1834 to 1839, 545 (60 percent) never completed their studies. The exact percentage of those who were expelled and those who quit for reasons of ill health or academic failure is unknown, but because school diplomas did not carry much weight in those days, most probably quit to take jobs.[24] John Weiss found that 61.4 percent of the students at the Ecole centrale during the same period did not finish their studies either.[25] It is likely that of the 60 percent of gadzarts who did not finish, 40 went to work, and 20 left for reasons of ill health, academic failure, or bad conduct.

The Second Empire

The first decade of the Second Empire, the 1850s, combined the worst of the barracks mentality of the First Empire with the religious obscurantism of the Restoration. The students continued to wear military uniforms and were governed by a book of minutely detailed rules and regulations. Military proctors supervised every movement from the beginning to the end of the day. The students were drummed awake at 5:30 A.M. and were subsequently drummed to the bathroom, to compulsory prayers, to meals (taken in silence), to classroom, workshop, and study hall, and finally to bed, always under the suspicious eyes of their proctors. The Sunday promenade was the sole outing of the week and was done in military formation. Students were never allowed any private moments, and their every movement was regulated and observed.[26]

Under the Second Empire each school had a resident Catholic chaplain. Sunday mass and morning prayer were compulsory during the 1850s but thereafter became voluntary. Le Brun, the inspector general, reported in 1866 that only a third of the Catholics at Châlons practiced their religion in any way, "much more than in the other two schools," where only one eighth and one tenth at Angers and Aix, respectively, "fulfilled their religious duties." The 40 or so Protestants at Aix and Châlons were only slightly more conscientious in their religious observances.[27]

With religion went punishment. Though corporal punishment was outlawed, a hierarchy of sanctions ranged from loss of outings, detention in the *salle de police* at the *table de pénitance*, to imprisonment in the school jail on bread and water and a bed of straw. After three prison terms in a year, the miscreant was expelled from school. Twelve types of violation were deemed so serious that they led to immediate expulsion; these ranged from the possession of a firearm to reading a newspaper or unauthorized book.

Students worked 12.5 hours per day, six days a week, and on Sundays and holidays they had 4 hours of study period. There were no holidays at all from the beginning of the school term in early October until Easter (eight days), and then none until summer vacation in August and September (eight weeks), which the students quite understandably referred to as *La Délivrance*.[28] They were not allowed to leave school for any reason, even for cases of serious illness, and could only be visited by their parents at the school in cases of emergency. Although the students were preparing to become industrial engineers, they were never taken to factories or industrial museums. As a graduate of Aix wrote in 1879 shortly after taking a job in Lille: "I spent three years at the school in Aix without ever having visited one plant, one industrial or art museum. I knew Lille better after one day than I knew Aix in thirty months."[29]

The Third Republic, 1879–1914

With the coming of the Third Republic in the late 1870s, the internal regime of the schools changed very little. Despite the professed liberalism of education officials, the schools continued to be petty police states. The school administration (the "Strass") was composed of a principal, an assistant principal, and an "engineer" who supervised technical programs, the shops, the procurement of materials, and the sale of goods manufactured in the shops. The bursar (*économe*) took care of financial administration. Under the Republic, the *censeur* replaced the chaplain and was placed alongside the engineer as assistant to the director. Stern and puritanical, he was in charge of discipline, which meant endless lectures on civic morality and republican orthodoxy.[30] Elaborate regulations governed every detal of the students' lives. The slightest lapse was observed and punished.

In the battle with the Strass the students had no allies. They might have turned to their professors, whom they respected when they were conscientious, but many of them, embittered by poor pay and the boredom of provincial life, were indifferent to the students. The good, younger ones were quickly wooed away by the better salaries and chances for advancement in industry. Those who remained at their posts were placed in a position that was clearly inferior to the administration. Although the faculty had some representation on the executive councils that ran the schools, they were outnumbered by administrators. They too were ranked in their own hierarchy: the professor of mechanics was on top, followed by the professors of mathematics, industrial design, and the liberal arts, and finally by the shop supervisors. Moreover, the professors had no clear place in the academic hierarchy of the French educational system; they belonged to no distinguishable corps. Despite the complexity of the courses they taught, they were not ranked at the level even of lycée professors, themselves badly paid.

Paul Nizan, the son and son-in-law of gadzarts, wrote in his novel *Antoine Bloyé* (1933) that "the professors look at the youth they instruct as though they were enemies."[31] Jules Guillou, an 1886 graduate of Angers, recollected that before World War I the professors were "used-up, embittered by boring work and the monotony of provincial life, feeling neither ambition nor joy."[32] The gadzarts had a song for the worst kind of professor they had to endure, whom they referred to as Professor V***:

Un chapeau Louis XI sur son chef enfoncé
Cache aux yeux du public son crâne tout rasé;
Il porte des lorgnons, mais sa vue est perçante,
Son regard est mauvais, sa voix forte et sifflante;
Il discute toujours, mais donne toujours tort.
Voilà l'insupportable, exécrable V***.[33]

Though an unpopular or incompetent professor occasionally provoked strife or even a riot, the real enemy was the Strass and their agents, the proctors. Like all other groups in the schools, the proctors had their own hierarchy, each individual reporting to a chief proctor, who had direct access to the principal and who was a member of the *Conseil de discipline.* In the lycées and collèges surveillance functions were filled by *maîtres d'études* or *répétiteurs,* who were either impecunious students working their way through school or *fruits secs,* unable for one reason or another to continue their studies but who at least knew the academic world. The proctors served no educational purpose in the schools; rather they were a cross between military policemen and prison guards. According to gadzarts, they were "tough but simple-minded old geezers, hardened by years of passive obedience and dried up by the colonial sun," and "brutalized by drinking too much absinthe." The gadzarts hated them for their cruelty and constant harassment, though some saw them as "more besotted than bad," their faces showing "more consternation than anger."[34] They too were victims of the system, and they were at least not righteous and puritanical like the Strass. Anyway, their malapropisms and non-sequiturs were an endless source of amusement: "When you have stopped being imbeciles, I'll begin," and, "If you've had that disease [typhoid fever] you either die or go crazy. I know all about it: I've had it."[35]

The proctors were called "pawns," "dogs," or "rats" by disrespectful students. They were stationed everywhere—the library, infirmary, workshops, classrooms, courtyard, study hall, bathrooms, and dormitory. Their tiny bedrooms, situated adjacent to the dormitory halls and dubbed "outhouses" (*chiottes*) in the school slang, were equipped with one-way peepholes for spying on their charges. The proctors even spied on the students in the toilets. Each one carried a notebook and reported anonymously to the head proctor, who could pass on the information to the director or censor without the student ever knowing about it. Marks for bad conduct were used to lower academic standing, so that the student never knew why he received the ranking he did.[36] A specified number of detentions led to expulsion. Hence the students were at the mercy of the administration.

Under both the Second Empire and Third Republic school officials were obsessed with what the age called "decency" (*la décence*), which was closely connected to a systematic denial of all bodily functions. This resulted in "la bataille des cabinets." A regulation of 1857 decreed that "in dressing and undressing students will observe the greatest decency, avoiding the slightest offense to morality; they must never go to the latrines, even at night, without their pants on."[37] Under the Republic a student could be punished for standing by the bathroom door, for making noise, wasting time, malingering, being untidy, and for "holding himself in an indecent position" in the bathroom. Never more than two students were allowed in these dangerous places at one time, a rule that caused little inconvenience at Angers, where there were only two toilets for the entire freshman class of 100.[38]

Despite their mania for "purity," the administration was remarkably indifferent to simple cleanliness. After spending six or seven hours per day in the oil and dirt of workshops and foundries, students were allowed to wash only their hands, face, and neck. Once a week they washed their feet, and once every three weeks they were permitted to shower. When the time for the shower finally came, the young men, aged 16 to 19, were marched by drums in order of rank to the dormitories. Then, under the watchful eye of the proctors, the following procedure was observed:

1. On arriving, the students will undress and will await the signal given by the drum in order to enter the water.

2. On the second roll of the drums, the students will come out of the water and get dressed. The proctors will assemble them into two ranks, marching them first to the dormitories where they will inspect their underwear, towels, etc. and then will march them to the dining hall.

3. In the bath, the students must not push or shove or knock down their comrades; under no condition may they go beyond the prescribed limits.[39]

Naturally the students did go beyond the prescribed limits whenever they could, as the following song indicates:

Couplets des Bains

Faut que j'vous parl' de not' sall' de bains
Qu'est très pratique,
Economique.
Tout est servi, ya pas d'turbin*,
A moins qu'on veuill' tourner l'robin†.
Tranquill'ment on r'tir' sa chaussure,
Accroche au mur
Toute sa nipure;
Et puis dam' l'on se met tout nu … u.
Les baignoir's n'dis't pas c'qu'ell's ont vu.

Refrain

Puis l'on tir' les rideaux,
Avant d'se mettre à l'eau,
Car sans ça, ling à ling,
ça s'rait pas comme y faut;
Et sans le faire exprès
On pourrait bien montrer
Son ling à ling à ling à ling à ling
A ling à ling à ling, à lingnote.[40]

On the recommendations of the principals and the inspector

---

*work
†robinet-tap

general, in 1873 the Commerce ministry completely separated the three divisions in each school. The students were not even allowed to talk to each other. The repeated protests of the alumni association were to no avail until finally, in 1897, the ministry partially lifted the restriction.[41] The last two decades of the century produced a series of director-martinets of whom Jacquemet, at Angers from 1879 to 1900, appears to have been the model. He was still incarcerating unruly students at the city jail, believing that the school prison, converted from the chapel of the old abbey, was insufficiently secure.[42]

In March 1900 a huge riot broke out in the school in which Jacquemet, the principal, and Jacquemart, the inspector general, were struck and injured by the students. The students had requested and been refused the abolition of the post of *censeur*, the suppression of the course in civic morality, the right to go out on their own on Sundays, and the privilege of a weekly shower. They barricaded themselves in the dormitory, holding out for two days before being evicted by military force. Fourteen students were expelled.[43]

Though relations had been generally good between administration, faculty, and students at Aix, by the 1890s the school had become a *chaudière* waiting to explode. In 1896 an uprising was caused by hazing practices that got out of hand. When the second-year students (*les pierrots*) broke all the rules by joining with the *anciens* to haze the freshmen, the freshmen resisted, uniting the first two divisions against them. The ensuing battle saw the conscripts so badly defeated that the principal had to evacuate the school. Here is a translation of his telegram to the minister of Industry and Commerce on 12 May 1896:

Very urgent. The prefect has forwarded your telegram to me. He and I agree that a proclamation of your appeal to the sense of responsibility and self-discipline of the students will have no practical result whatsoever given the extreme agitation here. This morning I evacuated the students of the third division in order to remove them altogether from mistreatment at the hands of the elders. Their panic is such that they refuse to return to the school this evening, and I have had to lodge them outside of Aix for the night. In these conditions, it seems absolutely necessary to suspend the first two divisions. The prefect agrees with me and has already made preparations for the evacuation of the school early tomorrow morning. Will send you a full report and recommendation for expulsion of the ringleaders.[44]

Two years later an even more serious riot occurred in which a student was killed (another had been killed in 1880 at Angers). The mutiny took several days to put down. The gadzarts long remembered "the events of '98" and sang a bitter song about the principal, whom they called "The Death-Dealer" (*Le Moricaud*):

Anniversaire du 10 Décembre

Le Moricaud dormait, lorsque sonna minuit;
Un fantôme tout blanc pénétra dans sa chambre.
D'une voix sépulcrale et se penchant vers lui:
"Réveille-toi," dit-il, "car c'est le dix Décembre!"

Souviens-toi de ce jour où tu fis tant de mal;
Ombre d'un Gadzarts mort, je descends pour te dire
Que nous te renions depuis ce jour fatal.
Spectre du châtiment, je reviens te maudire.

Souviens-toi, rongé par ce très vif souvenir.
Douze ans se sont passés, et c'est l'anniversaire;
Et si tu as du coeur, tu devrais en rougir.
Non! tu n'as pas agi en Gadzarts, ni en frère.

Va, nous te reléguons, car nous ne voulons pas
Admettre dans nos rangs un lâche et un parjure;
Si les Gadzarts vivants te renient ici-bas,
L'ombre d'un Gadzarts mort te crache à la figure.

Lorsque l'apparition se fut évanouie,
Le Moricaud, tremblant, entendit dans sa chambre
Une voix qui criait sur l'école endormie:
"Gadzarts! Souvenez-vous; car c'est le 10 Décembre!"[45]

For the gadzarts the cardinal sin of *lèse-solidarité* was expressed in the phrase "Tu n'as pas agi en Gadzarts, ni en frère." The principal was himself a gadzarts and therefore was branded a traitor, an outlaw for all time.

The punishment of the innocent caused many of the riots and rebellions in the schools. For example, in 1894 and 1898 at Cluny the principal made the following reports to the Ministry of Industry and Commerce on troubles within the school:

During the night of 28th November [1894], students of the second division threw projectiles at their night proctors . . . and broke the windows of their lodging. No student was formally recognized. Nine suspects were expelled. In January, 1895, one student, wrongfully expelled, was readmitted.

[In June 1898] a student in the third division thew a drinking glass at his proctor's head. The guilty one not having been recognized, two suspects were expelled.[46]

The pattern of revolts, or "mutinies" as they were called in the military jargon of the schools, was similar throughout the century.

They were a cross between a riot and a strike. Students gathered in quasi-military formation at the far end of the courtyard, boisterously singing their venerable war chants. They then ranged freely about the school, taking vengeance on any oppressor they happened to find. Finally, they barricaded themselves in a dormitory or workshop and began negotiations with school officials, though neither side had much to offer the other. So the director, dreading the reprimands and publicity that were certain to follow, called the prefect, who sent in the troops. The students, defying at once the school officials and the civilian and military authorities, held out for a time, "but under each window there shone a bayonet." Under siege "the students had no means of obtaining food or supplies and finally had to knuckle under. They filed out between two hedges of soldiers."[47]

During the 1880s troubles in the schools reached a crescendo; all four schools (Cluny was opened in 1891) remained in an almost perpetual state of revolt. There was even some minor sabotage in the workshops (*louper l'atelier, louper une pièce*: hence the word *loupins* for metal projectiles made clandestinely).[48] The Fontaine Commission of the alumni association reported in 1893 that between 1863 and 1884, 25 percent of students did not complete their studies (down from 60 percent during the 1830s), of whom 9 percent were expelled for misconduct, 7 percent for incapacity, and 9 percent left for various reasons (ill health, money problems).[49] In 1899 the Ministry of Industry and Commerce established a commission to study means of improving the moral and disciplinary regime in the Ecoles d'arts et métiers. It found that Angers had the highest rate of expulsions; Châlons followed closely behind, three times more than Aix. Indeed, if one counts only Angers and Châlons and allows for those who flunked out or left before they were thrown out, one can estimate that about 15 percent were obliged to leave the two schools for alleged behavior defects. From 1891 to 1899 the four schools produced 24 riots, several hundred expulsions, numerous injuries, and one death (Aix, 1898). Cluny, making up for lost time, led the pack with 6 riots and mutinies, and Aix, as always, trailed with only 2, though these were serious (1896 and 1898).[50]

Hazing Rituals

Gradually the student culture, with its songs, customs, esoteric language, ceremonies, rituals, and clandestine activities, took the form of an elaborately organized state within a state. This included an informal government headed by the *estime*, an elite of 10 student officers drawn from the senior class, whom the administration feared and sometimes even listened to.[51] The *bande noire* served as an internal police, enforcing a strict code of conformity among the students, intimidating proctors and even administrators, and creating formidable legends through their exploits. This accompanied quasi-religious observances, rites of passage and various ceremonies, a code of conduct, and an arcane language, all of which had to be taught to the conscripts in the face of the opposition of the Strass, which did everything it could to keep freshmen away from the elders.

The practice of hazing dated to the Restoration period, in which the students maintained a solid front against a persecuting administration. The latter expelled many students during the second half of the 1820s and then sought to separate new students from the remaining veterans. The veterans responded by enforcing a system of strict conformity upon the newcomers. Anyone who broke the rule of silence or in any way betrayed a comrade was forced to run the gauntlet (*roboteuse*) and to face ostracism from his fellows. Hazing thus became the crucial element in the formation of a student underground; it enforced the authority of the veterans over the conscripts and bonded newcomers to the collective personality of the group.[52]

When the new students, aged 15 to 16, arrived in early October, several days before the beginning of term, they were met by a cold and unsmiling Strass and the scowling *surveillants*. A few days later the elders arrived. Their first act was to march en masse into the freshman dormitory, singing boisterously:

Conscrit, t'en auras,
Conscrit, t'en auras,
Des coups de poing sur la gueule.[53]

This meant: "Conscripts, we'll get you ... with a fist in the kisser."

In an offshoot of La Rochefoucauld's old idea of *parrainage*, each conscript (called a beaver, *castor*, during the initiation period) was assigned a "godfather" according to his class rank: number 20 in the senior class was automatically the *parrain* of number 20 in the freshman class (his *filleul*). The elder promptly assigned his ward a *nom de guerre* (Ali Baba, the Strangler, and so on) and a notebook (*bucque*) in which he was to keep a record of his duties to his elders, confess his failures, and record his punishments.

For the first two months the godfather was at once the conscript's mentor, teaching him all the customs and traditions of the school, and his tormentor. The senior displayed his masculine qualities— his virility, toughness, and unconquerable will. And he must have seemed a formidable figure indeed to the freshmen. Aged 18 to 20, his military uniform worn in a rackish way, his hat slightly akilter, hands jauntily in his pockets, a large pipe in his mouth, slightly disheveled (*débraillé*), offhand and devil-may-care, exhibiting "un chic épatant, une crânerie inhérante à l'habit," always "un peu frondeur," he was in short "a superior being, telling tales of great exploits and heroic revolts, and expecting the new generation to do even better." To rise up to expectations, the conscript had to abandon his own will and acquire that of his superiors:

A ton ancien ne parleras
Qu'avec respect et poliment.
A ton ancien obéiras
Sans rouspéter* et promptement.
En entrant le salut feras
Et en sortant également.
Samedi soir 'basilique'[†] paieras
Et papiers-culs[x] également.
De tous les chiens, vesse[o] feras
Et de V*** également[#].
Noël, baptême recevras,
Si tu sais tes commandements.
Bon ancien tu ne deviendras
Qu'à ces conditions seulement.[54]

Most of the hazing involved pranks and stunts: singing funny songs (chanter les monômes) and dancing the *Zim-zim* and the *Cannibale* ("Conscrit!—une cannibale, ou je te pile les arpions!" "Dance the 'Cannibale,' or I'll stamp on your feet!"). Freshmen had to answer outlandish riddles and do improvisations on mechanical fantasies such as the following:

Raconter les péripéties d'une course de chevaux vapeurs attelés à des chariots de tour sur un champ magnétique.
Plombage de la dent du Midi avec le plomb du Cantal.

* without protesting
[†] collection for a worthy event
[x] paper wads
[o] beware, watch out
[#] this was the nickname given an unpopular professor

Examens à subir pour être nommé entrepreneur dc transports amoureux.
De l'antimilitarisme chez les infusoires.
Projet d'hospice d'aliénés pour poulies folles; préparation de la pâte denti-
  frice pour dents d'engrenage.
Du reboisement par arbres de transmission.[55]

Those who made the most baroque and outlandish (*poileux*) an-
swers won the greatest approval. Failure to respond well to interroga-
tion (*la fourchette*) led to punishments ranging from writing contrite
phrases 100 times on the four sides of a match stick to repeated kicks
in the backside (*cul botté*). The slightest disobedience was severely
punished. Though hazing had its lighter side, it could be severe and
even brutal, and had a serious purpose. As one graduate commented:
"La virilité des vieilles traditions se renforce: elles sont enseignées
durement: il faut aguerrir les conscrits."[56]

The hazing rituals of the early months were the first of a series of
ceremonies and festivals that continued throughout the academic
year. In the festival of Saint Eloi in December, conscripts changed
places with the elders for one day: until eight o'clock in the evening
they were allowed to haze the hazers. This was followed at Christmas-
time by a "baptism" ceremony in which the initiate (*castor*) was
cleansed of past inadequacies and accepted as a gadzarts worthy of
entering the brotherhood. He took the following oath:

Au nom de la République française, de la liberté, de l'égalité, de la fraternité,
et de la solidarité gadzarienne, j'abjure désormais le nom de castor pour
prendre celui moins méprisable de conscrit. Je jure sur le savonnier de mon
tabagnon, de respecter, de faire respecter, et de perpétuer les saintes traditions que m'ont léguées mes bons et vénérables anciens.[57]

After the new year a festival of reconciliation and assimilation
called *fignos* was held. This began as a conventional initiation ceremony, but at the sound of a trumpet all hazing stopped. Each
freshman walked between two rows of elders who tapped him with a
braided red cord called the *fignos*. The godfather then placed the
coiled braid into the lapel of his protégé, embracing him with the
words: "In the name of our old traditions, I dub thee gadzarts." This
was followed by a parade through town, the usual singing and
dancing in serpentine formation (*les monômes*), and an evening of
celebration: banquet, skits, music, dancing, and general hilarity.[58]

Fignos was followed in mid-February by *l'Equerre*, the halfway point
of the three-year term (508 of 1,015 school days), accompanied by

various celebrations. On *Ascension* the solidarity of each class was affirmed by smoking and passing the *pipe de promos* from student to student from all three *promotions*, followed by ceremonies and partying. The *Enterrement de Num's* was a mock funeral for the old school year, with students dressed in outlandish costumes. This was followed at the end of the school year by the biggest blowout of all, *La Délivrance*. Each school had its own way of celebrating this and other feast days, and each had its own particular *fêtes*. Aix called the year-end festival *La Grande Fusance*; Cluny, *La Grande Décale*.[59]

Language, Song, and Slang

The gadzarts had their own secret language. The hazing practices ensured rapid learning of the student *argot* and customs. Paul Gélineau made the first attempt to provide a lexicon of this slang in his book *Gadz'arts* (1910).[60] He compiled more than 500 words and expressions used in the five schools. In addition to the extensive vocabulary that built up over the years, there was something resembling an altered syntax. In commonly used compound words the first word was suppressed and replaced by z'à. For example:

z'à noire	—cravate noire
z'à dix, z'à seize	—paquets de tabac
z'à bleus	—pantalon bleu (worn in workshops)
z'à chic	—stylish, great, wow!

This combined with the habit of shortening words to produce such combinations as "z'à d'ép's" or *compas d'épaisseur*. Students frequently used shortened words such as *mécan's* for *mécanique* and *techno* for *technologie*. The gadzarts were also inventive in making up new words and expressions. The word for pass, move, or give became *F.C.* (*efse*) or *faire circuler*; hence "F.C. le couteau," "faire F.C. quelqu'un," and even "quand est-ce qu'on efse?" (when are you, or we, taking off?). Such usages varied according to the school and the time, although there has always been a basic vocabulary in common.

Most words in the student vocabulary had to do with the perpetual war with the *Strass*, or *Stration*, as the administration was also called. Enemy number one was the director, the *Jack's* or *pipin*, and the proctors (*chiens, rats, pions*), with their bad notes (*l'intégrale*) and their endless accusations and interrogations: *la fourchette* or *la planche* (*passer à la planche*). Normally the students found guilty of minor offenses were deprived of outings, for which there was an abundant vocabulary: *briqué, crampé, raclé, rousti de sortie*. The students' complaints and protests (*rouspètes*) were ignored, and all too many were expelled (*la sacquante*) without proper evidence. When the innocent were punished, the students screwed up their courage (*jus*) for the inevitable *chahut* or, variously, *chambard* or *boucan*. The signal for revolt was *le levée de pup's* (*pupitres*), the simultaneous banging of scores of desktops, which resounded throughout the schools. Once the revolt had been put down, the presumed ringleaders were invariably *sacqués par les Jack's*. These victims, who had been sacrificed for the common good, were *morts au champ d'honneur*.

Frequently those who were expelled were the better students, the natural leaders. One might even portray them as little Hectors, preferring *la sacquante*, to fall in the thick of the fray, rather than accept the inevitable surrender to the other side that they knew would come after graduation. Paul Gélineau argued that *la sacquante* was a form of death for the gadzarts. Three fourths of the students attended school on state scholarships; they were of modest origin, the hope of their families; and they had much to lose by expulsion or even the loss of their scholarships. To risk all this was a form of *amour-propre*, a way to maintain one's reputation in the brotherhood: "Just as one admires and loves those who brave death with insouciance, so one admires and loves those who play with *la sacquante*."[61] Or, to paraphrase Jesse Pitts, the leaders' defiance was an act of pure delinquent prowess, by its elegance and grace seducing the other students and uniting them around the cause for which they stood.[62] Though the leader would be expelled, *mort au champ d'honneur*, he would live on in the legends of the students, who, according to all observers, were great storytellers. The alumni association always accepted the *sacqués* into the society as true gadzarts and resisted all efforts of the Ministry of Commerce to limit the term *ancien élève* to those who had attained the school *brevet*. In 1899 17 members quit the association in protest when it accepted a candidate who allegedly refused to participate in a school uprising several years before.[63]

The students spent much of their time working in workshops. One word, deriving from the *atelier* and still in use, is *tabagnon* (*tabagon's* or *tabañon*). Originally a slang term for the office of the shop master, it came to mean the group of students organized around an assistant shop master (*sous chef d'atelier*), and his office, and finally to represent a basic unit of student organization. Each shop had its own name: the forge was *la chine*, and a smith *un chinois*. The foundry was the *lapinerie*, and a founder *un lapin*. The most prestigious shop was *ajustage* (fitting), in which the work was more skilled and clean than in foundry and forge, where one became "sale comme des gueux." In all of them one had to work (*bûcher*) very hard. The students were never really allowed to relax. Only for a time on Saturday night (*la pionce*) and again during the outing on Sunday afternoon could they enjoy themselves. No wonder that once in a while a student, worn out (*la flemme*), pretended to be ill so that he could spend a day or so "under the clean white sheets" of an infirmary bed:

On s'fait un petit air piteux,
On prend un'belle mine déconfite;
Et pour un jour ou bien pour deux
On s'cavale* à l'Infirmerie.[64]

Here one could "pique Larrens," that is, sleep late in the morning. In a school where the students were never allowed to sleep past 5:30 A.M., a mythical student named Larrens, who had managed to sleep until 6:30 one morning despite all efforts of the proctors to awaken him, was continually invoked as a symbol of resistance to the administration: "Chanter bien haut d'une voix triomphale; Larrens mourra quand mourront les Gadzarts." The cry, still heard in the schools, of "Vive Larrens" invokes the venerable spirit of solidarity among gadzarts.[65]

The students were capable of extraordinary daring. Perhaps their most spectacular feat came at Cluny on the night of 14 July 1897, Bastille Day, when five members of the *bande noire* placed the black flag of the band and the tricolor on top of the medieval tower of the old monastery of Cluny, one of the tallest in France at 65 meters. They hid inside the tower during the evening and, when all was quiet, ascended to the top. One of them then climbed out on a ledge, slithered up the lightning rod to the roof of the tower, and attached the flags to a large cross and weather vane that had been placed there many years before. The perch was so perilous that no roofer in the area could be found to take down the flags.[66]

Marxist critics have accused the gadzarts of various kind of capitalist failings, including misogyny,[67] but there is little evidence to support this claim. The students doubtless had their underground world of dirty songs and limericks, and they did a lot of masculine strutting, as schoolboys do, but of the few references to females in their pub-

---

* take off, split, go

lished songs and traditions, most involve nostalgia for home, family, and sweetheart, from whom the young men had been so rudely separated. An occasional sexist remark was made in speeches at alumni association banquets, especially when an old boy could not resist the temptation to give advice to young graduates: "The choice of a career is as delicate as the choice of a woman; you have to try several before you find a good one."[68]

After 1900, when the students could leave school and go into town at times on weekends, they were able to pay court to the opposite sex. The *viscrits* or *pierrots* (Aix), second-year students exempt from the time-consuming hazing process, were supposed to specialize in *chambards* and chasing women, leaving the responsibility of properly educating the *conscrits* to the *anciens*. The only piece that deals completely with the problem of women is not unsympathetic in tone:

Ça Manque de Femmes chez les Gadzarts

Puis la nuit on fait de beaux rêves,
On se sent nager dans l'bonheur;
Mais c'est pas vrai, et quand on s'lève
On en a toujours gros su l'coeur.
Chaque matin, c'est un supplice
Quand il faut sortir du plumard*.
Va falloir qu'j'écrive au Ministre
Qu'ça manque d'femmes chez les Gadzarts.[69]

On the whole the students seems to have been normal young men decidedly more interested in the joy of Bacchus than in the moralistic admonitions of the Strass:

Couplet de Bacchus

Loin de ces murs de si triste mémoire,
Enivrons-nous en de joyeux festins;
Du Dieu Bacchus, fêtons ici la gloire,
Et loin de nous les soucis, les chagrins.
Censeur, cessez votre morale austère,
Car sachez bien qu'à l'Ecole des Arts,
Ce jus divin qui réjouit la terre
N'emplit jamais le verre d'un Gadzarts.[70]

Such was student distrust of the administration that even the liberalization of 1900 aroused their suspicions. They saw in the new

---

*bed

regime a more subtle, but just as pernicious, form of social control:

Le Nouveau Règlement

I

Je suis simplement
Et tout particulièrement
Chargé de vous éduquer
Et de vous inculquer
Les plus beaux sentiments.
Grâce à mes leçons
Mes soins et ma bonne façon,
Je ne désespère pas
De vous remettre au pas
D'une bonne éducation.
Pour vous, plus d'argot,
De jurons ni de gros mots.
Nous rognerons l'aile à toutes les traditions;
Bien vite démoli
Sera le mauvais esprit
Qui vous pousse encore à la rébellion.

II

Je veux arriver
A complètement transformer
L'ensemble de nos travaux
Par des moyens nouveaux
Dont tous seront satisfaits.
Plus de surveillants
Pour vous punir; mais cependant
Tout ce qu'ici vous ferez
Me sera rapporté;
Oh! fort discrètement.
Et dans mes bons jours,
Faisant patte de velours,
Je vous accorderai un peu de liberté.
Oui, par la douceur,
Je veux gagner vos coeurs;
Je serai sûr alors, sur vous, de pouvoir régner.[71]

Nevertheless, combined with the efforts of the alumni association and the ministry, the liberalization gradually improved the atmosphere in the schools. Students could come and go more freely, engage in sports and outdoor activities, and visit factories and museums. Food was more plentiful and of better quality, and classes were smaller and more informal. Although the old dislike of the Strass died slowly, and antics, pranks, and even occasionally riotous behavior continued to occur, there were no more mutinies.[72] The more extreme forms of conduct disappeared after the turbulence of the 1890s. The gadzarts were in the process of becoming engineers and were expected to behave themselves. During the twentieth century the spirit of resistance was gradually converted into an esprit de corps in imitation of the Ecoles centrale and polytechnique. How this took place, and the role of the alumni association in the transformation of the *Esprit Gadzarts,* will be discussed in the following chapter.

The Ecoles d'Arts et Métiers: How Others Saw Them, 1880–1980

Vous qui cherchez ce bonheur qu'on méprise
Rendez-vous donc à l'école des Arts,
Egalité, c'est là notre devise.
C'est la devise de tous les vrais gadz'arts.
Traditional song of the Arts et Métiers

In the previous chapter we discussed the life of the schools during the nineteenth century, notably the traditions of resistance, clandestine self-government, and elaborate rites and rituals developed by the students over the course of the century. In this chapter we will discuss the gadzarts' social ideas and attitudes—how they saw themselves and how others saw them. In order to portray their collective mentality and social ideas, we shall examine their songs and sayings and the reports of officials and outside observers. The alumni association played an important role in defining and codifying the *Esprit Gadzarts* and in representing it to the outside world as part of its campaign to improve student behavior, create a better public image, and win the engineering diploma for the schools.

The reports of the school principals and inspectors on the internal situation of the schools, though numerous, are surprisingly devoid of understanding of the roots of disciplinary problems among the young men. School officials were apparently more concerned with impressing their superiors with their own moral qualities and administrative capacities than with providing an accurate picture of what was going on in the schools and the students' state of mind. Most ruled by the book, enforcing every detail of the regulations, and were seemingly caught off guard by the resultant student explosions. During the nineteenth century none could explain why the students behaved the way they did; hence they had few suggestions as to how to solve the problems. J. A. Vincent, director at Châlons during the 1830s and inspector general for a decade thereafter, blamed the "rejects" from the collèges, especially those of Paris, for the disciplinary problems. These "déclassés" were less manageable than the working-class boys, who were at least grateful for the chance to receive an education.[1]

Vincent's successor at Châlons, N. A. Le Brun, knew the schools as well as any official during the century. A polytechnicien, he was shop master at Châlons in 1839, assistant principal at Angers in 1843, principal at Châlons from 1848 to 1855, and then inspector general of the schools to 1870. In his annual report for 1855 he admitted that there was "a bad spirit and the germ of insubordination" among the students but that "this seldom carries over into their working lives after leaving the school." He attributed the disciplinary problems at the schools to three causes: "First, I would emphasize the poor family environment and training so common today in all classes but particularly in the lower ones.... Secondly, the smirking Parisians who amuse themselves by involving their fellow students in the worst kind of silliness.... Thirdly, the strong feelings of camaraderie and solidarity among the students ... who do not recoil before the frequent expulsions of students from the schools" and who manage in any case to find jobs in the mechanical industries, "which are anxious to hire our students."[2]

Le Brun had no suggestions for improving the situation beyond the usual policy of repression and emulation. He suggested taking away scholarships from rebels; he removed a copy of Montaigne's *Essays* from the library shelves as "too suggestive" and fruitlessly tried to persuade the students to go to mass, though never more than a third did so. He fell back on the old formula of emulation, recommending the hiring of principals and professors who "by their distinguished bearing and education can at once impress themselves on the students and serve as examples in such a manner as to soften, if possible, the rough manners and crude upbringing of the great majority."[3] No wonder the schools vegetated, the faculty became embittered, and the students furious. In 30 years as principal and inspector, Le Brun could think of nothing better to do than remove a copy of Montaigne from the library shelves.

The incomprehension of school officials and their ignorance of adolescent psychology forced the students into opposition. Not until 1904 does one find a principal's report, that of M. Corre at Lille, revealing an understanding of the psychology of adolescent groups. He observed that the students "are individually neither better nor worse than other young people their age, having bad heads and good hearts." The problem lay in the group mentality:

Here the individual personality is effaced, blending into a highly emotional and unpredictable collectivity in which the individual remains anonymous and his acts impersonal, escaping all direct responsibility and all effective sanction.... Add to that the memory of a semi-cloistered regime, the military discipline that ruled the arts et métiers for almost a century, and the force of tradition of which the basic feature is a deep-seated almost instinctive distrust of the Administration ... and you can understand why we have problems in the schools.[4]

During the 1890s it therefore became increasingly clear that the boarding school regime had to be either thoroughly reformed or abolished. The Fontaine Commission of the alumni association reported in 1893 that the turbulence in the schools was disrupting studies and that the students being expelled were often the better ones.[5] This caused Jules Siegfried, the minister of Commerce, to name an official commission to investigate the causes of the problems in the schools. In its report of 1897, *rapporteur* Louis Bouquet, who was also the director of the technical education division, contrasted "the brilliant positions attained by graduates in various industries at home and abroad" with the "collective acts of indiscipline so frequent in recent years."[6] But he was unable to explain why discipline had broken down so completely, especially when the schools corresponded so well, at least in theory, to Plato's idea of the harmonious development of mind and body: "He [Plato] would look with pleasure upon the vigorous bodies of our young workers, their solid muscles developed through healthy activity and hard work. The buildings are large and well-aired and the cities in which they are located are healthier than most." Yet, "indiscipline is permanent in the study hall: they [the students] spend their time plotting against their proctors; they sing, they stamp their feet, and they neglect their assignments. If it weren't for their fear of being asked questions in class, the young men would do nothing at all except in the workshops. There one can still find intense activity and joy in work." The commission concluded that "the students need to have healthy outlets for their energies, to play games, to run and to make noise.... It is especially important to extinguish in their hearts those morose sentiments and that bitter hatred of authority which is evident everywhere."[7]

In 1902 an article by Emile Hinzelin appeared in the *Manuel Général de l'Instruction Publique*, the organ of the Ministry of Public Instruction, which gave a detailed account of the 25 riots and mutinies in the four schools during the 1890s. The article was based on a confidential report in the files of the Ministry of Industry and Commerce of 7 April 1899, which had been leaked to the rival Ministry of Public Instruction. Why, asked Hinzelin, were the Arts et Métiers the source of so much more trouble than the other schools? He referred to "the unfortunate vestiges of the quasi-military organization" and to the abandonment of the old practical programs and objectives. A key explanation lay in the nature of the students themselves: "Do not forget that at the same time as they are schoolboys, they are workers who are thoroughly versed in the art of handling iron and wood. A good part of their day is spent in the workshop where many acquire an incredible skill and a legendary capacity for work. With two turns of a file they can make a key that will open any door in the building."

Obviously the fact that gadzarts were workers as well as students could explain much of their resistance to authority, but Hinzelin could not explain why the disciplinary situation was deteriorating so badly during the 1890s: "Their behavior seems to have become more barbarous during the past few years. Their slang is becoming harsher and more vulgar. Cynicism replaces the picturesque. Hazing especially has taken on a repugnant, even murderous, character. Certain students, tormented by their fellows, have been crippled for life."[8] Ironically, the Ecoles d'arts et métiers had been thoroughly reformed in 1899, and the source of some of the tensions had been removed. Perhaps the Ministry of Public Instruction, still smarting in 1902 from the loss of the Ecoles nationales professionnelles in 1900 to the Ministry of Industry and Commerce, obtained a certain satisfaction in publishing confidential files reflecting unfavorably on the schools under Commerce.

Nevertheless, there were enough incidents in the schools, even after the liberalization of 1899, to cause concern in the Ministry of Commerce. In 1908 Gabelle, director of the technical education division, said: "We must recognize that the Ecoles nationales d'arts et métiers have a rather peculiar spirit, which is a source of worry sometimes to all who are interested in their future."[9] That same year the ministry almost abolished the *internat* after a series of troubles in the schools. In 1912 Paris was opened as an *externat* over the protest of the alumni association, which declared that its graduates "can't be true gadzarts."[10]

Within industrial circles the Ecoles d'arts et métiers were widely praised for their contribution to French industry and the arms race with Germany.[11] But outside, few writers and intellectuals mentioned them, and those who did were seldom favorable. Echoing Arago, Corbon, and Le Play, Paul Nizan criticized them during the interwar period for uprooting the sons of workers from home and workplace and placing them in an artificial boarding school environment. Nizan's novel, *Antoine Bloyé* (1933), was based on the life of his father, Pierre, a graduate of Angers in 1883.[12] Pierre had spent 46 years working for the Western Railway Company before his death in 1930. The son of a railwayman and the grandson of Breton peasants, he had been educated on state scholarships in the enseignement spécial program in a lycée and then in the Ecole d'arts et métiers at Angers. He married the daughter of a fellow gadzarts, Paul Métour (Angers, 1859), depot master for the Orleans Railroad Company in Brittany.[13] Thus Paul Nizan was the son and son-in-law of gadzarts.

In the novel Antoine Bloyé rises higher than most gadzarts employed in railroad companies. He begins his career as a fitter for the Orleans Company and becomes an engine driver, depot master, and then plant manager in various cities of western France. Early in the twentieth century he is promoted to the coveted position of chief engineer and managing director of a large plant manufacturing railroad equipment, supervising thousands of workers and employees, the *bâton du maréchal* for a man of humble origin working for the railroad. During the Great War the plant is converted into an armaments factory producing explosive shells. When some of these shells prove defective, Bloyé is made the scapegoat and demoted to a position as head of stores ("a mere green-grocer"). His retirement follows soon after, and it is a despairing one. Having spent his life in constant activity, he is ill-prepared for leisure and reflection. He begins to realize that his life has been a failure; he has sold out his class, his true love, Marcelle (a working-class woman), and ultimately himself for the "cotton wool of bourgeois life." He has passed to the side of the bosses:

All his efforts, all his memories, altered not one jot of his complicity. He thought of his father, who was one of those who took orders, of his comrades in the shipyards of the Loire and in the railway depots who were also on the side of those who serve, on the side of life without hope. And returning home in the icy Auvergne dawn, he repeated to himself a phrase that held good for the whole of his life, a phrase that he forced himself to forget, that only disappeared in order to reappear in the time of his adversity, on the eve of his own death: "So, I am a traitor."[14]

In retirement Antoine Bloyé engages in various activities to give his life meaning, none of which works. He joins the local chapter of the alumni association but soon comes to realize that the banquets, balls, lunches, and benefits only disguise the fact that "no profound human end united these men." For a time it is good to recall the memories of schooldays, the antics, the pranks, the offbeat songs, the romanticized adventures. But soon:

In the stuffy Marguery salon, amid the plaster ornaments, the wrought-iron lamps, the rails, the glass, the potted palms, men who had aged stirred the memories of their schooldays with a sort of lofty melancholy.... But by the time the cheese came, these men, separated by social distances, by the families they had raised, by the variety of their professions, the different marks they had made in the world, with their inequality of manners and success, had little left to say to each other.... They became again middle-aged and elderly men weighed down by the sauce-covered meats and over-rich cooking.[15]

This "conjuring of shadows" cannot disguise from Bloyé that he has become no more than "a fragment of a man, alienated, mutilated, a stranger to himself."[16]

The real Pierre Nizan may not have been burdened by such remorse in his old age. He appears to have successfully contributed to the war effort: in 1920 he was promoted to chief engineer of a large district in Alsace, the province newly returned to France from Germany, where he directed the reorganization of the railway system for five years before retiring. During the 1920s his son, Paul, received the finest education possible in France, graduating third in his class from the elite professors' training school, the Ecole normale supérieure, just behind classmates Jean-Paul Sartre and Simone de Beauvoir.[17] Had Pierre Nizan written an autobiography, he might have reached different conclusions, for his family had risen in four generations from illiterate peasants and workers to engineers and prominent writers and moralists.

How, then, did gadzarts see themselves and the outside world? The most persistent themes were based on their perception of themselves and the workers as common victims of the bureaucrats, their feelings of loneliness and isolation in the schools and the business world, their faith in hard work as a means of building a better future for themselves and for France, and their sense of fraternity and solidarity with each other in the face of a hostile world.

The idea of gadzarts and workers as common victims of the bureaucrats in charge of French industry and government, so well expounded in the 1870s and 1880s by Denis Poulot, lessened their sense of guilt over the betrayal of their class. If producers of all types were victims of *fonctionnarisme* and the unfair competition and incompetent management of the polytechniciens and other *pantouflards* from the civil service, they must cease to fight each other and band together in the patriotic duty of freeing French industry from the hands of bumbling bureaucrats.

Some gadzarts emphasized their teaching function as middle-level supervisors situated between the bosses and the workers. In the early decades of industrialization the gadzarts played an important role in the large firms in instructing workers, many of whom were newly arrived from the countryside, in using machine tools, reading machine drawings, and adapting to the rhythms of the factory. As workers began to organize unions, the gadzarts, as the production engineers who had to deal with them, became more hostile, though they distinguished between the unions and the workers themselves.[18]

The gadzarts were undoubtedly very hard workers, as the motto of the alumni association—"Work, Order, Probity"—suggests. They believed that the workers should follow their example of advancement through education and work. Otherwise they tended to avoid social questions in their discussions. Nizan's portrait of Antoine Bloyé as a man so devoted to his job that he has no time for ideas, or even for his family, may be an accurate one.

The gadzarts were neither poets nor writers; they had little or no instruction in language or the liberal arts. Once employed, they seldom had time to read. Hence their social ideas were limited to the expression of vaguely Saint-Simonian ideas and free enterprise myths. Denying the existence of class conflict, they talked of progress and humanity and were quite ready to excuse any variety of evils— poor working conditions, low salaries, industrial pollution—in their name:

S'il obscurcit les cieux de longs flots de fumée,
Ce n'est pas une insulte à leur sérénité:
Assis, pensifs, auprès de la forge allumée,
Dans ses rêves, ils voient grandir l'humanité.[19]

There was, however, a darker side to this optimism. The gadzarts was aware that he was uprooted, an outsider in the system, struggling to get ahead against difficult odds. His sense of isolation began with the awareness that he came from a pariah class and had been taken out of his milieu and placed in a barracks/monastery environment designed to remold him along alien lines:

Il me fallut quitter mes parents, mon amie.
Sous les cloîtres maudits, aux murailles noircies,
Gadz'arts, j'allais passer trois ans dans les ennuis.
Angers, pour trente mois, remplace mon pays.[20]

Monastic and prison themes frequently recurred in student songs as symbols of loneliness and isolation. At Angers, where the prison was a converted chapel, the two themes came together:

Là, dans un coin qu'était tout noir,
Sur un'vieill' planch' fallait s'asseoir;
Avec pour vivr' jusqu'au lend'main,
Une cruch' d'eau, un morceau d'pain.
Logés de cett' façon civile,
Y avait pas l'temps de s'faire de bile.
Ah! ah! c'est épatant!
C'qu'on était chouette au bon vieux temps (*bis*).[21]

In the ancient monasteries and convents in which the schools were located, the "cloître silencieux" conjured up ghostly images and even a vision of death:

Jadis, dans ton ombre, cachée,
J'ai prié le coeur en émoi.
Jadis, tournant sous les arcades
Avec les nonnes à pas lents,
Durant leur morne promenade,
Je rêvais d'amour en pleurant.
J'aimais un joyeux mousquetaire;
Mais mon coeur n'avait pas d'espoir,
Et je suis morte solitaire
Dans le couvent aux froids couloirs.[22]

Very few gadzarts were religious, and insofar as they were political, they embraced a nonideological, secular, republican individualism. Most of them married, frequently the daughters of other gadzarts.[23] Their lives were apparently focused on work and competition rather than on their families. And yet at work they had little social contact with their colleagues from other schools; the lack of sociability of junior managers in business and industry forced the young men to look to their alumni associations for human solidarity and professional contacts.[24] And yet, as Antoine Bloyé found, "no profound human ends united these men." They were quite alone in the world. As one graduate said, "La vie est un combat." Deeds, not words, counted; one could only continue the struggle in the service of progress, stoically and without complaint.

Vous savez pourtant que la vie
Est réellement pcu de chose,
Et qu'un rien a souvent suffi
Pour que tout se métamorphose.
Soyez donc sérieux; et songez
Que ce n'est que pour peu de temps
Que vous êtes là. Travaillez,
Espérez et soyez patients.[25]

The students had a strong sense of social injustice and a firm belief in the values of liberty, equality, and fraternity. But these beliefs tended to turn inward, to action within the alumni association, rather than outward into public life. Few gadzarts became involved in politics or community affairs, and despite their modest origins, even fewer became active in socialist causes. Of those who did, the most notable were François Jeandeau (Châlons, 1828), deputy in 1848, Eugène Tartaret (Châlons, 1837), a member of the International during the Second Empire, and Arthur Groussier (Angers, 1878), a socialist deputy and later vice-president of the Chamber of Deputies from 1893 into the 1920s.[26]

The social mentality of the gadzarts can be illustrated by the discussions of a study group of young graduates meeting in Boulogne-sur-Seine in 1918 just after the Great War. The group complained that it was much harder for the sons of workers to make it to an Ecole d'arts et métiers than for the sons of the bourgeoisie to prepare for the *baccalauréat* in mathematics:

The affluent classes have a wide variety of schools and careers available to them, while the lower orders have only the Ecoles d'arts et métiers open to their elites. The disproportion is flagrant; on one side the great mass of working people with one small outlet for their best minds, on the other a limited number of fortunate ones who have all the possibilities in the world at their disposal.[27]

The group complained that graduates of the Ecoles d'arts et métiers were widely considered to be "subaltern engineers," yet they were clearly more capable and productive on the job than graduates of the great schools:

They [the gadzarts] have not received the recognition that their efforts, and their contribution to the war, deserved.... [Because] they suffer from an inferior social status, they are placed at the mercy of the graduates of the great schools ... who are not truly competent for the posts they hold.

Thus France was moving even closer to the creation of "a new administrative and industrial feudality."

The only proposal that the young graduates could come up with was a moratorium on the establishment of new Ecoles d'arts et métiers, "for the creation of too many schools will undermine our diploma.... Our elders acquired their credential by the force of their own hands, by hard work and merit. Thus it is our property." Seeing the alumni association as the embodiment of the gadzarts' spirit and the protector of their achievements, the Boulogne group called upon it to play a more active role in the protection of their *positions acquises*. Denunciation of the system and implied threat of revolt gave way in real life to a defense of vested interests and of the diploma as a piece of property. The students' solidarity thus continued to turn inward after graduation in an associational life that was defensive rather than outward looking.

The example of the Boulogne group combined with themes of isolation in student songs and solidarity against the Strass provide an excellent illustration of the Pitts-Wylie-Crozier model of authority relationships in France.[28] This model stresses the dislike of face-to-face confrontation in French organizations and the tendency to take refuge in impersonal codes of behavior. As noted, the code governing the Arts et Métiers was modeled on the Napoleonic barracks and the early factory. The assumption underlying the model was that young workers would not willingly labor from dawn to dark for the better part of three years without being forced to do so, and that they would not, as future foremen and supervisors, give unconditional loyalty to their bosses without strict coercion and control. Once the authoritarian model was established, the Strass had no power to alter it. They could only enforce the detailed rules and regulations issued by the Ministry of Industry and Commerce, which became more rigid as time went on. The crucial decision-making power concerning student conduct had been transferred upward to the ministry in Paris, though such officials were in a very poor position to legislate rules of behavior over schools located hundreds of kilometers away. As Michel Crozier said: "A bureaucratic organization is an organization that cannot correct its behavior by learning from its errors."[29]

This explains why the regime of *casernement* lasted without reform for almost a century. What characterized the life of the Ecoles d'arts et métiers, in contrast to other boarding schools (which had their share of troubles but to a lesser degree), was that there were so few areas of flexibility or, in Crozier's words, "areas of uncertainty," in which negotiations could take place. There were no mutually shared goals or esprit de corps among the students, the administration, the alumni association, and the ministry.

During the first half of the century, however, there had been some room for maneuver. La Rochefoucauld had tried to improve behavior through a mixture of persuasion, philanthropy, and manipulation, with some success. His successor, Boisset, tried to divide and rule at Châlons by governing through a select group of students, but he failed miserably in the face of student hostility to his tactics and religious principles. Dauban successfully introduced a humane regime at Angers during the July Monarchy, which worked well for 16 years but fell afoul of the disorders of 1848. He was made the scapegoat and was retired early. His was the fate of all those who took responsibility and were caught out. The failure of the principals by mid-century to solve the problems of indiscipline led to a transfer of power to the Ministry of Industry and Commerce. Under the Second Empire the cautious and unimaginative Le Brun became the model of the successful academic administrator. He applied the rules even-handedly, never took any initiative, and always deferred to higher authority. He retired in 1870 with the usual honors and pensions after 35 years of faithful service.

The transfer of final authority to the ministry in Paris left the Strass with little freedom of maneuver, while the students became ever more intransigent. The result was a system of parallel levels of authority that governed the schools. The students accepted the authority of the Strass as legitimate only insofar as it provided the knowledge necessary to professional success. But in questions of conduct they recognized only their peer group, as represented by the elders of the *Estime* and as enforced by the *bande noire*. Thus the senior students could exercise strict control over the freshmen through elaborate hazing rituals. Despite the increasing tendency of these rituals to get out of hand and result in injury, the alumni association had no choice but to defend hazing and all other traditions. Its own raison d'être as the incarnation of the gadzarts' spirit demanded no less than total solidarity with the students. The old boys defended hazing because "in order to learn to command one must first learn to obey," but of course such strictures did not always apply to the Strass.[30] This created a very delicate situation for the association; by always backing the students, no matter how bad their conduct, the executive offended officials in the schools and the ministry, thus hindering its efforts to improve relations in key places.

As the situation within the schools worsened toward the end of the century, there were some murmurs of dissent within the association. In 1884, after a student had been killed at Angers and several others seriously injured as a result of a series of riots that had taken place over the previous years, two members of the association's executive argued in the annual general meeting that the organization should openly cooperate with officials in repressing hazing practices. They even dared to defend the ministry's policy of separating the three divisions, the abolition of which was the association's highest priority. After a long and very heated debate, they received the support of only one member in an assembly of well over one hundred. Their proposal was so soundly defeated that it was a long time before anyone dared speak openly again in favor of such a policy.[31]

The students knew that the association would back them to the hilt. Anyone *sacqué par les Jack's, mort au champ d'honneur*, received the full support of the association's placement bureau in finding a position, and because industry was always looking for skilled men with or without diplomas, those expelled had little trouble finding jobs. The *sacqué* was much better off than the *mouchard* who cooperated with the administration; the latter was henceforth an outsider, ostracized from all contacts with gadzarts in the association and in industry. This is why even innocent students accepted expulsion rather than break the code of silence. The principals, knowing full well that they often expelled the innocent, knew also that if they did not make examples, they might themselves be *sacqués* by a vengeful administration in Paris.

Given the deadlock in the schools, one wonders why the Commerce ministry did not convert the schools into *externats*, transferring them to industrial centers where they could take advantage of the educational and scientific facilities of the universities. The Commerce ministry was reluctant to do so because of the expense involved and because affiliation with the science faculties meant cooperation with the rival Ministry of Public Instruction. The alumni association was opposed to any such project, as were the cities of Aix, Angers, Châlons, and Cluny. Moreover, Commerce officials and many industrialists believed that the boarding school was the best way to produce socially reliable managers and technicians. The nineteenth-century mind was obsessed by the threat of disorder. Many of the students came from the working classes, and some had even participated in revolutionary agitation in 1815, 1830, 1848, and 1871. The industrial army would not be secure without trustworthy lieutenants, and thus the young gadzarts had to be made into examples of order, discipline, and intelligence to the workers, not ringleaders against the established order.

Until the end of the century, therefore, the instinct of officials had been to tighten, not liberalize, the disciplinary regime in response to disorder; hence the separation of the three divisions from 1873 to 1897. But this policy led to a standoff in the schools by the 1890s. At Aix the students seriously believed that their principal was responsible for the death of one of their comrades (*le Moricaud*), and the principal at Angers regularly put recalcitrant students in the city jail. Châlons and Cluny were the scenes of chronic disorder. The collapse of discipline and the breakdown of the Napoleonic system at least restored an element of flexibility to the situation and opened the way to reform for the first time in decades.

The alumni association had already decided that liberalization was the only policy acceptable to all parties. In 1897 it obtained a partial end to the separation of divisions. In that same year the Commerce ministry created a Conseil de perfectionnement for the school and granted the association 15 of 23 seats. This gave it much more leverage in both the ministry and the direction of the schools. Then, in 1899, Alexandre Millerand became minister of Industry and Commerce. He was one of the rare ministers who combined strength of character with a relatively secure tenure in office and a strong government behind him. And he was willing to make changes. As he told the alumni association banquet: "Before coming to the schools these young men are generally well-behaved and hard working. . . . After graduation they go into industry and the military services and everyone speaks very well of them. It is only at the schools that they are turbulent."[32]

The alumni association replied that such turbulence derived not from student disloyalty but from the harsh internal regime introduced by the Restoration and its abandonment of the liberal and humane ideas of the Duke de La Rochefoucauld. The key to reform in the schools lay in the return to the duke's ideas of the education of the whole person and of work in common. In response to criticism within the association and to reluctant Commerce ministry officials that liberalization might lead to anarchy, the association argued that opening up the boarding school regime and awarding the engineering credential were the only alternatives to continuing disorder.

Such arguments turned the tide in a Republic that, at least in theory, revered the memory of the duke and his liberal principles. Accordingly, in October 1899 Millerand abolished the military regime: the proctors, the uniforms, and the strict regimentation. He introduced informal classes, sporting and leisure activities, and more frequent outings and leaves. This was followed by a strengthening of admission requirements and programs. After 1901 every candidate to the schools had to possess a diploma from the Ecoles nationales professionnelles, the Ecoles pratiques d'industrie, or the Ecoles primaires supérieures.[33] This completed the process by which the Ecoles d'arts et métiers were promoted to the level of secondary technical schools, which opened the way to the engineering credential.

In return for so many concessions the association apparently made a tacit agreement with Millerand that it would exert more pressure on the students to behave themselves. This it did, placing before them the inducement of an engineering credential for dignified comportment.[34] It also quietly sought to appropriate the student culture, the "Esprit Gadzarts," by claiming to represent and embody that culture on an ongoing basis. Just after the turn of the century the association established a Comité de Traditions, which organized similar committees within the schools. Together these committees set about codifying student customs. The result was the *Carnet de Traditions* of 150 pages, a kind of student catechism and Bible that still may not be seen by outsiders. In 1910 Paul Gélineau published a book on school life that revealed many of the sacred rites and rituals, and this was followed in 1925 by a similar book by Alfred Metton.[35]

In defining and codifying the gadzarts' spirit, the association shrewdly emphasized the traditions that reflected the gadzarts' deep need for religion and family. It constantly repeated such phrases as "saintes traditions," "le feu sacré des traditions," "le carnet de famille," "fêtes de famille," "la grande famille des gadzarts," combined with continual reference to such slogans as "fraternité, voilà notre devise," and "notre solidarité, unique au monde." Calling each other "camarade," they reassured themselves that they belonged to a true fraternity of equals regardless of rank and position obtained in life. This solidarity had been forged in their struggle against the Strass and in their underground governments, initiation rituals, tests of courage, and sacred oaths. The graduates were persuaded that in the past they had risked everything (*la sacquante*) and stood the test; hence they were not traitors to their class at all. They had proved themselves.

The Duke de La Rochefoucauld was central to all themes. He was the common father in a brotherhood in which women were only a romanticized and shadowy presence, at least before the advent of a few gadzarettes beginning in the 1960s. He was the protector and benefactor of noble blood, unjustly persecuted for his good works. He had died despised by the Restoration regime and his body profaned, but his spirit lived on as the prophet of a golden age of opportunity in which the producer, not the bureaucrat, would be justly rewarded. The religious symbolism of the good and suffering father, who had been an outcast during the Revolution and an outsider during the Restoration, and who therefore understood what they were going through, was a powerful one for young men of modest origin. During the first generation, many pupils had been war orphans, and later, cut off from home and family, they were at least symbolically so. Thus the duke provided the gadzarts with a sense of family, a vision of a better future, and an aura of gentility that they needed so badly to share.

The alumni association was the Church, the visible presence of the Esprit Gadzarts, erecting monuments to the Duke de La Rochefoucauld on the old estate at Liancourt, the place of pilgrimage for all gadzarts. The founder's memory lived on thanks to his devoted disciples, who dispensed gratitude like holy water. As one association officer said in 1900: "The memory of that great personality is etched ineradicably in the hearts of all graduates of the schools of Arts and Trades. As a sign of gratitude, they never fail to evoke his memory in their meetings and friendly gatherings."[36] The association purchased scores of busts of the duke, which were given to each local group. At every banquet the bust, enrobed in the crests of the schools and bedecked with various flags, memorabilia, and relics, was placed in a central part of the hall as a kind of shrine. It was *de rigeur* to toast the duke as if he were still alive and among them. Hippolyte Fontaine's toast to the duke in the Paris banquet of 1885 is typical in tone and stilted style:

When the hour for speeches has sounded and the President rises to propose a toast to the founder of our dear schools, we always join together in remembering the Duke de La Rochefoucauld-Liancourt. The memory evoked is so intense in all generations of students, so alive among us, that our wine glasses ring to the repeated cries of: 'To the health of the Duke de La

Rochefoucauld-Liancourt,' even though he has been separated from us now for fifty-eight years.

A century ago many industries were enveloped in the darkness of professional ignorance, and when La Rochefoucauld-Liancourt founded the first Ecole d'Arts et Métiers one can say without exaggeration that he built a beacon destined to light the way of pioneers in the mechanical industries and that he endowed France with one of her finest and most useful institutions.[37]

Fontaine's toast was followed, according to the report, by "several salvos of bravos testifying at once to the constant and inalterable gratitude in the hearts of all graduates for the one who has made them everything that they are in the world, and the noble sentiments of gratitude that they profess for all the true benefactors of humanity."

During the years before World War I, the association sent its officers to orient incoming freshmen in the nature of the gadzarts' spirit and in appropriate conduct, before they arrived and fell totally under the influence of the senior students and the administration. These "chats" (*causeries*), as they were called, sent curiously mixed messages to the conscripts; the Strass was still the enemy but no longer the enemy it had once been; antics, pranks, and hazing were part of "les trads," but the future engineer was expected to behave with a certain dignity and bearing (*tenue*). It was important to project a better image to the outside world, and rioting, excessive hazing, and messy hair were hardly the stuff of which good public relations were made. One association officer told a group of incoming freshmen in 1910 that "the alumni association absolutely and energetically opposes all disorder or any behavior contrary to good discipline; in this it is entirely in accord with the Minister of Industry and Commerce who will continue to deal with any infringement of the rules and regulations in a manner that is inflexible but just." But then, for fear of appearing to side with the Strass, he hedged his point: "Unfortunately the direction [of the schools] sometimes, in good faith, punishes the innocent whom it mistakes for the guilty." He concluded by exhorting the students to dress well, be less unkempt, and speak French more correctly.[38]

The effect of all this was not so much to alter behavior or impair the esprit de corps as to modify them just enough to prevent the extremes that had resulted in such bad publicity for the schools. The hazing rituals and the old "trads," now codified and suitably trivialized, combined with sports and other robust pursuits to fashion a team spirit that was coming into vogue in the modern business corporation. Hence the alumni association gradually transformed the traditions of the Ecoles d'arts et métiers in a manner that was the exact inverse of their original intention: the students' solidarity against the Strass became an esprit de corps for advancement of position within the system; traditions of revolt became traditions of acceptance; and oral traditions became written ones, thus losing their force. An oppressive military regime was liberalized without sacrificing the principle of manipulation and control. Some students and the association members recognized and resisted this process, but they could do little in the face of apparently benign changes conducive to the general advancement of the schools and their graduates.

The gradual improvement in the schools' atmosphere after 1900 was interrupted by the Great War, which saw the loss of a fourth to a third of the prewar classes. One survivor of the *promotion* of 1912 recalled his return to school in 1919: "You no longer saw big laughing boys who overwhelmed an incomprehending administration with a thousand vexations; now you saw only men, their faces lined and careworn, etched in the pain of four years of war. There were the sick and the crippled, seeking to recapture a lost springtime, a springtime that they had never known and never would know." Despite all this, one graduate recalled, "We were not granted the least liberties to which our age and experience entitled us. The administration simply could not give up the old military system."[39]

Another graduate noted how little the reigning mentality in the schools had been changed by the new psychology and permissiveness of the interwar period: "The same old forbidding lecture halls, the same glacial dormitories, the grimy workshops, the blind and brutal authoritarianism of the administration which treats young men as unruly children, even those matured beyond their years by the ordeal of war." And so "the struggle between students and administration went on, each side choosing its own weapons." The administration was, as always, "suspicious and punishing, expelling without pity and for the slightest reasons." The students in turn became hardened, and when the tension became too great, "they exploded in enormous riots, the echoes of which went well beyond the walls of the schools. The B.N. [*bande noire*] organized raids and reprisals, . . . and the struggle to win over the conscripts increased." In this "detestable atmosphere" the students "somehow managed to keep their humor and a certain gaiety, thanks mainly to their traditions; the long chain which has been forged for over a century is always strong."[40]

Despite the tenacious tradition of resistance, most observers (Gélineau, Metton, Pillot, and Popin) have agreed that the atmosphere in the schools has slowly improved since the turn of the century.[41] In 1947 Paris became the central institution, grouping all students from the regional centers into a common fourth year. The fact that it was an *externat* alarmed the association, which proceeded to build a dormitory in the *Cité universitaire* at its own cost. The government showed its appreciation by building a freeway, the Boulevard Périphérique, right in front of it. For a time after the war there were few incidents, but the 1960s brought a decade of *contestation* and a renewed challenge to the association from the students and the younger *promotions*. After a serious incident at Châlons in 1964, the government abolished the compulsory *internat*.[42] That same year a seventh campus was opened at Bordeaux-Talence as an *externat*, and the old traditions penetrated slowly.

The union of students (Union des élèves de l'ENSAM), founded in 1950, backs the Société des Ingénieurs Arts et Métiers wholeheartedly in advancing the interests of ENSAM as a *grande école* and in opening wide the channels of professional advancement.[43] To achieve these ends, they defend the *internat* and the old "trads" as a means of building school spirit in competition with other *grandes écoles*. Since the interwar period the students have taken to calling hazing and initiation rituals *usinage*, which suggests that the process is one of machine tooling and refining the freshman raw material into a uniform product for business and industry. This is stated in the preamble to the *Carnet de Traditions*:

It is necessary to forge your personality, to mould your spirit, to temper your character. The possessive adjective does not exist for you. As the president of the association never tires of saying, you must acquire a certain kind of personality and an authentically Gadzarts spirit. The purpose of our traditions, he says, is to create from the outset a single type of man, avoiding all deviation in behavior.[44]

Both the Société des Ingénieurs Arts et Métiers and the Union des Elèves put considerable pressure on freshmen to submit to *usinage*. About a tenth refuse to do so and are disdainfully dubbed *H.U.* (*hors usinage*). They do not fully take part in the celebrations and festivals that mark the school year and that carry over after graduation into a network of contacts in the association and in business and industry.[45] The two organizations never tire of warning that the "trads," and therefore the brotherhood of gadzarts, are fragile: the end of the compulsory *internat*, the existence of the *H.U.* outside the band, and the appearance of gadzarettes in the school (never spoken but often implied) all threaten the old ways. In fact, however, as a reflection of a popular student culture, the traditions have long since been dead. The old resistance of working-class boys to forced deculturation and reacculturation in bourgeois values has been transformed into an esprit de corps facilitating the absorption of students into the corporate structure.

Gadzarts still do not reflect much on the social role of the engineer. Recently they have emphasized the responsibility of the engineer as the man who designs and builds the structures of the modern world, whose projects have a great impact on the human community, and who must keep abreast of the dizzying advances in the technology of machines and information. The association and the union of students argue that gadzarts should acquire, as leaders of men, something of a *culture générale*, and especially a greater fluency in written and spoken French, but they do not specify how students can accomplish this while taking so many science courses.[46]

Pierre Pillot (Lille, 1923), secretary-general of the alumni association from 1937 until his retirement in 1976, has argued that during the nineteenth century fraternity and solidarity replaced Christian charity as the ideal of conduct in the West. We serve others through common work and through our solidarity with them rather than through charity directed at them. During their 200 years, the Ecoles d'arts et métiers evolved as a microenvironment of the human struggle for liberation: the useful work done in common, the good humor and fraternity in the face of oppression, and the sense of the real in dealing with intellectual and human challenges. According to Pillot, the fraternity of gadzarts is a model for others to follow in modern times but only as it opens out into the world. The greatest danger is that it will degenerate into an *esprit maison*, into clannishness, becoming just another public relations weapon in the competition with other higher technical schools.[47]

Pillot's successor, Pierre Gaudin, has moved the society in the direction of a purely professional association seeking to advance the schools' status through various political and professional activities.[48] The executive continues to advance the position of the Ecole nationale supérieure d'arts et métiers in the hierarchy of the technical *grandes écoles* in France. It seeks to maintain links with the past mainly by encouraging the old "trads" and residence in the *internat*, which have long since lost their original meaning and vitality.

Today the average student in ENSAM is no longer cut off from the outside world by enforced residency in the schools. He may come and go as he pleases, live where he wishes, and spend periods of time working in industry. The main problem is no longer for the schools to get in touch with the outside world as for the outside world to get in touch with the schools. Aside from Bordeaux and Paris, which are surrounded by other institutions of higher learning and scientific research, the centers remain in relative isolation in lesser provincial cities. They have highly sophisticated equipment and research facilities, but these are not always shared with other schools and researchers. Most current research involves contractual projects of a specific nature for industry. Although this is all to the good, ENSAM continues to some extent to exist in a world apart, outside the mainstream of French scientific research and culture.

Social Origins and Careers

Dans la vie, comme dans la navigation à voile,
il y a des forces positives et des forces contraires.
L'homme doit utiliser même les forces contraires; le
drame, c'est quand il n'y a pas de vent.
Alfred Sauvy

In this chapter we will study the social origins, careers, and fields of work of more than 2,000 graduates of the Ecoles d'arts et métiers during a century and a half of industrialization in France. Specifically, we will present: (1) a study of the social origins of 661 graduates from the period 1810–1890 and a similar study of the parents and grandparents of all 600 graduates of the class of 1978; (2) an analysis of the careers of 1,370 graduates of the *promotion* (first year of studies) from 1820 to 1920, tracing last positions down to the 1950s and 1960s; and (3) a series of studies, taken from the database of 1,370 graduates, indicating first jobs, the number of years for advancement, and places and areas of employment. Though gadzarts did not enjoy the professional opportunities of their social superiors from the Ecole polytechnique and Ecole centrale, the statistical studies of their professional achievements presented in this chapter indicate that they were remarkably successful considering their humble origins and the modest programs and goals of the schools.

The main sources for the studies of social origins and careers were (1) the school files in the F 12 series of the Ministry of Industry and Commerce in the *Archives Nationales*; (2) the alumni association's *annuaires*, which provided the occupation, firm, place of work, and residence of the members; and (3) the association's special editions of the *annuaire*, entitled *Listes générales*, which were published in 1900 and 1923 and provided professional histories and residences on all graduates, dead or alive.[1] The *Listes générales* included both members and nonmembers, although the information provided on members was much more extensive. Obituary notices in the association's *Bulletin administratif* were also consulted when appropriate.

The files and publications of the alumni association of the Ecoles d'arts et métiers constitute our main source of evidence. Founded in 1846 and originally confined mainly to the Paris area, the association grew rapidly after 1860. In 1880 30 percent of graduates were members; in 1900, 60 percent; and in 1930, almost 90 percent. The experience of living together in the *internat* built up a strong camaraderie among the students, which fostered the continuation of contacts thereafter, even to the point of numerous business partnerships and family alliances. As membership fees were not high, the main determinant in the decision to join seems to have been the loyalties and antipathies built up in the schools rather than professional position and wealth. Well endowed by large gifts and bequests from wealthy alumni, with a budget of 1 million francs by 1900, the association provided extensive social, mutual aid, and placement services that attracted a broad cross section of graduates.[2]

Professions of Fathers: The Nineteenth Century

Table 10.1 presents the occupations of the fathers of a sample of 661 graduates from the classes of 1810 to 1890, divided into three periods: 1810–1830, 1830–1860, and 1860–1890. Table 10.2 presents a three-generational study of the fathers and sons of a subsample of 127 graduates over roughly the same period. As one can see by examining the occupations of their fathers in table 10.1, most graduates from the period 1810–1830 were from the public, military, and professional sector and were mainly the sons of small government employees (27 percent) and soldiers (23 percent). As might be expected, they showed little aptitude for technical studies and seldom ended up in industry. The period 1830–1860 coincided with the coming of the railroads, steam technology, and the mechanization of industry. The reform of 1832 placed the Ecoles d'arts et métiers under the Ministry of Industry and Commerce and required a year's apprenticeship of all candidates, thus considerably reducing the percentage of the sons of petty officials and soldiers (12 percent and 8 percent, respectively). The rapid rise in industrial demand for graduates beginning in the late 1830s won over a solid clientele of artisans, skilled workers, and small factory and shop owners who came mainly from the machine and metal trades. Creation of the Ecoles primaires supérieures by the law on public education of 1833 also contributed to improving the quality of incoming students.

In comparing the period 1810–1830 with that of 1830–1860, we note an increase in the sons of artisans (17 percent to 27 percent), most of whom were metal workers and smiths, and in the sons of manufacturers, managers, and businessmen (10 percent to 17 percent). In industry the number of factory foremen and workers increased (2 percent to 10 percent), while those in the public, military, and professional category declined steeply (52 percent to 22 percent).

Between 1860 and 1890 the public, military, and professional sector fell further to 13 percent. The sons of those in the artisanal, small business, and farming category dropped by more than one half (41 percent to 18 percent). However, the sons of those in industry almost doubled (33 percent versus 67 percent), with the greatest increase being in the percentage of foremen and factory workers (10 percent to 27 percent), a reflection of industrial growth and the decline of the old artisanal trades. The workers were mainly fitters, machinists, mechanics, and engine drivers. Taken together with the artisans, they comprised 37 percent of the fathers from 1830 to 1860 and a similar percentage (38 percent) from 1860 to 1890. The percentage of working people thus remained large and constant over the century, but the composition changed, artisans and smiths gradually giving way to skilled factory workers.

From the first period (1810–1830) in table 10.1 to the last (1860–1890), the sons of manufacturers, businessmen, engineers, and supervisors increased from 12 to 37 percent of the sample. Toward the end of the last period, the number of independent owners within this group began to decline and that of salaried engineers, technicians, and employees to increase, a trend that continued in the twentieth century as the age of free enterprise gave way to industrial concentration. By 1890 two thirds of the fathers (67 percent) were in the industrial categories (manufacturers, businessmen, engineers, foremen, and skilled workers), and almost one fifth (18 percent) were in the artisan, small business, and farming category, mainly artisans. The line between artisan and worker was rather difficult to draw because both groups were associated with the machine and metal industries and trades (there were few butchers, bakers, and candlestick makers among them). Generally the gadzarts were of modest origin. Although they were seldom poor, they were closely tied to the world of work and production.

The line between prosperous artisans and small manufacturers in

Table 10.1
Occupations of fathers of 661 graduates of the Ecoles d'arts et métiers, 1810–1890

	1810–1830		1830–1860		1860–1890	
	(n = 262)	percent	(n = 244)	percent	(n = 155)	percent
Industry						
Manufacturers, managers, and businessmen	25	10	42	17	32	21
Engineers	0	0	3	1	12	8
Supervisors and technicians	5	2	8	3	12	8
Employees and draftsmen	5	2	4	2	5	3
Foremen and workers	5	2	25	10	42	27
Subtotal	40	16	82	33	103	67
Small Business and Farming						
Artisans	45	17	67	27	17	11
Small business and shopkeepers	19	7	18	7	4	3
Farmers	18	7	18	7	6	4
Subtotal	82	31	103	41	27	18
Public, Military, Professional						
Liberal professions and high officials	4	1	0	0	0	0
Employees and clerks	70	27	30	12	6	4
Teachers	4	1	5	2	8	5
Military	60	23	20	8	6	4
Subtotal	138	52	55	22	20	13
Miscellaneous	2	1	4	2	5	3

NOTES: In the industry category businessmen were mainly contractors in public works and railway construction. There were relatively few *négociants*. Among farmers, most were *cultivateurs* and *agriculteurs* rather than *propriétaires ruraux*, who were usually better off: between 1810 and 1830, 10 were *cultivateurs* (4 percent) and 8 *propriétaires* (3 percent); between 1830 and 1860, 11 were *cultivateurs* (4 percent) and 7 *propriétaires* (3 percent); and between 1860 and 1890, 5 (3 percent) versus 1 (1 percent). In teaching, 1810–1830, 2 were secondary and technical school professors and 2 primary instructors; between 1830 and 1860, 1 and 4, and between 1860 and 1890, 1 and 7. In the military category, 1810–1830, 28 were lesser officers (11 percent), and 32 were noncommissioned officers and enlisted men (12 percent); between 1830 and 1860 9 were officers (4 percent), and 11 were NCOs and enlisted men (4 percent); between 1860 and 1890, 2 (1 percent) and 4 (3 percent).

Table 10.2
Occupations of the fathers and sons of 127 graduates of the Ecoles d'arts et métiers, a three-generational study

	1st-Generation (Fathers)		2nd-Generation (Graduates)		3rd-Generation (Sons and Sons-in-law)	
	(n = 69)	percent	(n = 127)	percent	(n = 201)	percent
Industry						
Manufacturers, managers, and businessmen	20	29	68	53	61	30
Engineers	1	1	31	24	81	40
Supervisors and technicians	1	1	11	9	6	3
Employees and draftsmen	2	3	0	—	0	—
Foremen and workers	12	17	2	2	0	—
Subtotal	36	51	112	88	148	73
Small Business and Farming						
Artisans	14	20	4	3	0	—
Small business and shopkeepers	2	3	0	—	0	—
Farmers	6	9	2	2	2	1
Subtotal	22	32	6	5	2	1
Public, Military, Professional						
Liberal professions and high officials	2	3	2	2	27	13
Employees and clerks	3	4.3	0	—	1	1
Teachers	3	4.3	5	4	6	3
Military	3	4.3	2	2	12	6

Subtotal	11	16	9	8	46	23
Miscellaneous	0	—	0	—	5	2

NOTES: Among farmers in column 1, all 6 were *cultivateurs* (9 percent); in column 2 both were *propriétaires*; in column 3 both again were *propriétaires*. Among teachers in column 1, all 3 were primary instructors, and in columns 2 and 3, all were secondary or technical school professors. In the military category all were petty officers and NCOs.

the machine and metal trades was also difficult to define, for there were many small artisanal firms in France having one to five employees. In his study of 721 graduates taken from the registers of the Ecoles d'arts et métiers from the years 1860–1875, Michel Bouillé found that 15 percent had come from the upper classes [propriétaires-rentiers (5 percent), liberal professions (1.4 percent), manufacturers and businessmen (6.6 percent), and higher civil servants (2 percent)].[3] In my study of 155 graduates from 1860 to 1890 (table 10.1, column 3), I found that 21 percent were the sons of manufacturers and businessmen (which included propriétaires who owned businesses) and 8 percent each of engineers and technicians. However, I found only 14 percent artisans and shopkeepers, as opposed to Bouillé's 25 percent. As for workers (27 percent to Bouillé's 22 percent and farmers (4 percent to 6.6 percent), we obtained similar results. I probably defined as small manufacturers certain small producers whom Bouillé called artisans. Many industrialists, for example, used artisanal titles such as "maître de forges" (Petin, for example) or "serrurier-mécanicien" (Cadiat). In consulting obituary notices, I found that fathers' professions changed over time; a father might be a serrurier (metal smith) when his son was a student and a serrurier-mécanicien (an independent manufacturer) some years later. I chose the last professional entry when I found more than one. When so many firms were artisanal in nature, the dividing line was obviously hard to discern, and this (and the small number of subjects in this particular study) may explain the difference between my findings and those of Bouillé.

A Three-Generational Study

Table 10.2 traces the occupations of the fathers, sons, and sons-in-law of 127 graduates of the Ecoles d'arts et métiers, most of whom finished their studies between 1830 and 1870. This detailed biographical information derives from an analysis of the obituary notices in the alumni association's *Bulletin administratif* and thus probably represents a more successful group than would be obtained by a random sampling. In the first generation (the fathers of the graduates) slightly less than one third (31 percent) were the sons of manufacturers, managers, businessmen, engineers, and supervisors, and more than one third were the sons of artisans and workers (20 percent and 17 percent, respectively). Only a minority (16 percent) came from the public, military, and professional sector. In comparing these figures to the general sample of table 10.1, 1830–1860, we find that a higher percentage of first-generation graduates in table 10.2 had fathers who were manufacturers, managers, and engineers (30 percent versus 18 percent) and that fewer came from the public, military, and professional sector (16 percent versus 22 percent). The percentage of sons of foremen, workers, and artisans was the same in each table (37 percent). These data suggest that the sons of manufacturers, managers, and engineers were most likely to succeed; following that it was best to be the son of a skilled worker or artisan and worst to be the offspring of a soldier, schoolmaster, petty employee, or shopkeeper.

Members of the second generation, the 127 graduates, were also more successful than average. A large majority reached high-level positions in industry as manufacturers, managers, and businessmen (53 percent) or as engineers (24 percent). The question of the educational and occupational choices of their 201 sons and sons-in-law, the third generation, is fascinating, for they faced no financial or geographical obstacles to higher education and the choice of a profession. One would expect them to have chosen the liberal professions and higher civil service in a country reputed to hold industrial careers in low esteem. Generally, however, the sons and sons-in-law (column 3) followed in the footsteps of their fathers. Nearly three fourths went into business, industry, and transport, mostly as business executives or as owners or partners in their own firms (30 percent) or as engineers (40 percent). About one fourth were employed outside of business and industry, mainly in the liberal professions (13 percent), as professors in higher, secondary, and technical schools (3 percent), or as superior military officers (6 percent). Eighty-eight percent of their fathers worked in business and industry, and 73 percent of the sons and sons-in-law continued in the same fields. Given the variety of available options, the fact that so many followed after their fathers argues against the traditional theory, which holds that industrial groups were constantly decapitated through the desertion of sons into the professions and government services.

It was also possible to trace the education of one half (100/201) of the sons and sons-in-law in the third generation. The great majority (79 percent) were enrolled in technical and business schools; of these, more than one half (52 percent) attended an École d'arts et métiers, demonstrating once again the very close ties that the gadzarts and their families had with the schools and with industry. The remaining sons and sons-in-law graduated from the Ecole centrale (12 percent), the Ecole polytechnique (10 percent), law and medical schools (10 percent), other technical and business schools (5 percent), Saint-Cyr and Beaux-Arts (6 percent), or were still studying in lycées (5 percent).

The Twentieth Century: The Class of 1978

Thus far we have discussed social origins during the nineteenth century; table 10.3 brings us to the present day. It is based on a study done by the Société des Ingénieurs Arts et Métiers on the occupations of parents and grandparents of 600 students in the class of 1978 of Paris.[4] This was the last class to graduate under the regime introduced in 1945 (first graduating class in 1949), which advanced the Ecoles d'arts et métiers to university status and added a fourth year of studies. The students, aged 21 or 22, were born around 1956 or 1957. Their fathers and mothers (table 10.3, columns 1 and 2) were born around 1930, and the paternal and maternal grandparents (columns 3 and 4) during the early years of this century. The paternal and maternal grandparents came mainly from the traditional artisanal, shopkeeping, and farming sectors and less from industry (45 percent versus 39 percent for the paternal grandfathers and 51 percent versus 34 percent for the maternal grandfathers), while the fathers (column 1) had already made the move to industry (62 percent versus 25 percent involved in the small business, artisanal, and farming sectors). Only a small percentage (13 percent or less) in any generation came from the public, military, and professional category.

During the twentieth century the independent owners—manufacturers, businessmen, artisans, and shopkeepers—were gradually replaced by salaried executives and employees. The paternal grandfathers were more likely to have been independent owners than were the fathers of the students in the class of 1978, many of whom were salaried executives, managers, and engineers (23 percent), or company employees (12 percent) and technicians (15 percent). The percentage of sons of foremen and workers (12 percent) had declined gradually since 1890, when it had stood at 27 percent (table 10.1), while company employees increased from 3 percent in 1890 to 12 percent of the fathers in table 10.3.

The mothers of the class of 1978 (table 10.3, column 2), married during the 1950s, were probably the last generation of homemakers that we will see in this century. Seventy percent were "sans profession," and those who worked fell almost entirely into occupational categories traditionally associated with women: secretaries, clerks, and teachers. The maternal and paternal grandparents (columns 3 and 4) were mainly farmers (32 percent and 27 percent) and artisans and small shopkeepers (19 percent and 18 percent). They generally came from families in the traditional small business and farming communities (51 percent and 45 percent), although a fair number were already involved in industry (34 percent and 39 percent), usually as workers and employees. As always, few came from the public sector. During the twentieth as well as the nineteenth century, the schools fulfilled their prupose of providing educational opportunities and industrial careers for young people from modest families, most of whom were involved in industrial, commercial, and agricultural pursuits.

The 1961 Survey

Table 10.4 is based on a selection of business and technical schools from a 1961 survey on the occupations of fathers of French *grandes écoles*: the Ecole polytechnique, the Ecole centrale, 12 Ecoles supérieures de commerce, and the Ecole nationale supérieure d'arts et métiers.[5] The Ecoles d'arts et métiers had fewer sons of manufacturers, businessmen, and executives and more sons of workers, artisans, small industrial employees, and technicians than each of the other, more prestigious, schools. Perhaps surprisingly, the sons of businessmen and manufacturers preferred the Ecoles supérieures de commerce over the Ecole centrale, while the Ecole polytechnique recruited heavily from the liberal professions, the higher civil service, and the professorate.

In comparing the results of the 1961 survey with the alumni association's study of 1978 on the Ecole nationale supérieure d'arts et métiers (tables 10.3–10.5), we find that the categories do not perfectly coincide. The 1961 survey (table 10.4, col. 4) states that 4 percent of the students at the Arts et Métiers were the sons of manufacturers, 6 percent of businessmen, and 8 percent of executive employees. The 1978 study (table 10.3) did not specify private owners and manufacturers, referring instead to the categories of *président-directeur-général* (*PDG*), *directeur-général* (*DG*), and vice-president, the most prestigious corporate posts (1 percent), followed by the *cadres supérieurs*, executive engineers and managers (22 percent). The 1961 survey (column 4) suggests that about 10 percent of the fathers of gad-

Table 10.3
Occupations of parents and grandparents of 600 Students, ENSAM Class of 1978

Profession	Fathers		Mothers		Paternal Grandfathers		Maternal Grandfathers	
	(n = 600)	percent	(n = 600)	percent	(n = 600)	percent	(n = 599)	percent
Industry								
President, V.P.	5	1	3	0.4	13	2	0	—
Engineers	133	22	5	1	21	4	21	4
Supervisors and technicians	88	15	21	4	27	4	13	2
Employees	75	12	72	12	69	12	67	11
Foremen and workers	69	12	16	3	101	17	102	17
Subtotal	370	62	117	20	231	39	203	34
Small Business and Farming								
Artisans	26	4	0	—	45	8	61	10
Small business and shopkeepers	40	7	16	3	59	10	56	9
Farmers	84	14	0	0	168	27	189	32
Subtotal	150	25	16	3	272	45	306	51
Public, Military, Professional								
Liberal professions and high officials	24	4	5	1	27	5	27	4
Teachers	37	6	40	7	14	2	10	2

Military	16	3	0	0	18	3	5	1
Subtotal	77	13	45	8	59	10	42	7
No profession	3	—	422	70	3	—	5	1
No reply	0	—	0	—	35	6	43	7

SOURCE: Based on a study conducted by the Société des Ingénieurs Arts et Métiers in 1978.
NOTES: Among farmers in column 1, 63 (11 percent) were described as small and as medium; in column 3, 147 (24 percent), and in column 4, 165 (28 percent). Among teachers in column 1, 29 (5 percent) were higher, secondary, and technical professors, 8 (1 percent) primary teachers; in column 2, 11 (2 percent) and 29 (5 percent); in column 3, 3 (0 percent) and 11 (2 percent); in column 4, 5 (1 percent) and 5 (1 percent). No designations were provided for the military columns.

Table 10.4
Occupations of fathers in selected grandes écoles, 1961

Father's occupation and percentage in 1954 occupational census	Polytechnique percent	Centrale percent	Commerce percent	Arts et Métiers percent
Industry, Business, Agriculture				
Manufacturers (0.4)	5	3	12	4
Businessmen (7.7)	6	7	17	6
Executives (1.0)	14	20	17	8
Middle employees and supervisors (2.8)	3	7	6	8
Lesser employees (10.9)	8	9	5	10
Workers (38.9)	2	2	5	17
Artisans (3.9)	2	2	3	9
Farmers and farm workers (26.8)	1	2	4	6
Public, Military, Professional				
Liberal professions (0.6)	16	7	8	3
High officials (0.9)	19	16	8	6
Professors (secondary and university) (0.4)	8	4	1	2
Teachers (2.0)	9	5	2	5
Lesser government (1.1)	3	6	6	6
Others	3	4	1	5
No profession, retired	1	6	5	5

SOURCE: *Les Conditions de développement, de recrutement, de fonctionnement et de localisation des grandes écoles en France*. Rapport du Groupe d'études au Premier Ministre, 26 Septembre 1963. Paris: La Documentation Française, 1964. Cited by Fritz Ringer, *Education and Society*, p. 196.

zarts were manufacturers and businessmen, while the 1978 study says that 1 percent were. Perhaps these terms no longer necessarily signify independent owners but rather certain types of higher executives. The general conclusions for the top industrial categories (manufacturers and executives) of the two tables correspond fairly well: 18 percent in 1961 versus 23 percent in 1978 (tables 10.4 and 10.3, respectively).

There are a few other problems in the definitions of categories. The 1961 survey did not distinguish shopkeepers and petty commerce, which were probably mixed in with "business" and with artisanal crafts and shops. The 1978 study did not distinguish between public and private employees. The number of farmers was considerably higher in 1978 than in 1961. Nevertheless, the two studies confirm the trends already apparent in the late nineteenth century toward salaried employees from business and industry and away from independent owners, large and small, and from the public, military, and professional sector (table 10.5).

In table 10.5 comparing the 1969 and 1978 surveys, we find that more than one fourth of gadzarts came from the bourgeoisie (manufacturers, liberal professions), and more than one half were of modest origin (artisans, workers, employees, and so on). The number of technicians and middle-level supervisors grew from 8 to 15 percent between 1961 and 1978. Clearly the Ecoles d'arts et métiers have continued to draw students from modest social groups. The first class under the *grande école* regime, introduced in 1974 and implemented in the schools in 1976, graduated in 1979, so it is too soon to know whether the social composition of the Ecoles d'arts et métiers is changing. About 20 percent of the students are admitted through *passerelles*, from professional programs and institutions. Most of these are trained technicians possessing various certificates and diplomas from the lycées d'enseignement professionnel, continuing education programs, the Instituts universitaires de technologie, and the Conservatoire national des arts et métiers. But most of the students still come from the *classes préparatoires* of the lycées, general and technical. The Ecoles d'arts et métiers may end up becoming indistinguishable from other *grandes écoles* in their recruitment, most of which draw from a fifth to a fourth of their students from modest social groups.

Career Patterns

In the following pages we will trace the professional advancement of 1,370 graduates of the Ecoles d'arts et métiers from the classes of 1820

Table 10.5
Comparative table on the occupation of fathers of students of the Ecoles d'arts et métiers, nineteenth and twentieth centuries

	Classes 1830–1890 percent	1961 Survey percent	Class of 1978 percent
Industry			
Manufacturers, managers, engineers	22	18	23
Technicians and middle supervisors	5	8	15
Employees and draftsmen	2	10	12
Foremen and workers	17	17	12
Subtotal	46	53	62
Small Business and Farming			
Artisans and shopkeepers	27	9	11
Farmers	6	6	14
Subtotal	33	15	25
Public, Military, Professional			
Liberal professions and high officials	0	9	4
Employees and clerks	9	6	—
Teachers:			
secondary and technical	1	2	5
primary	3	5	1
Military	6	3	3
Subtotal	19	25	13
Miscellaneous	2	7	—

SOURCE: Table 10.1 columns 2–3 (n = 399); Table 10.4, column 4; and Table 10.3, column 1.

to 1920. In the case of twentieth-century graduates (up to the class of 1920), careers will be traced through the 1950s and into the 1960s, the time of retirement. We will begin by discussing the first positions held and then follow with a series of studies of last positions, defined as the last jobs held at least 29 years later. We will discuss the time sequence for career advancement, showing how long it took to move to each professional level. Finally, we will indicate the industries in which gadzarts worked and where they were located. The purpose of these studies is to measure the professional achievements of gadzarts in industry and transport and in the government services, to demonstrate which fields they preferred, how well they did, and how their preferences and prospects for promotion changed over time.

First Positions

Table 10.6 illustrates the first position held by graduates of the Ecoles d'arts et métiers. Originally (1815–1830) about one fourth (23 percent) of the graduates were employed in the public and military services, especially in the Ponts et Chaussées (15 percent), but this figure declined thereafter. Most graduates during the early years began their careers as artisans and workers (59 percent). In the following period (1830–1860) the number of skilled workers (43 percent) and draftsmen (28 percent) increased rapidly. Generally the graduates followed one of two overlapping paths. During the middle decades of the century they usually started out as machinists, mechanics, and foremen, becoming shop supervisors and then plant and production managers. As the century progressed an increasing number began as draftsmen in planning departments (*bureaux d'études*) and sought to advance to positions as department supervisors (*sous chef et chef d'études*) and finally to director of research and development (*directeur technique*). As their responsibilities grew, they acquired the title of engineer and frequently became plant, divisional, and technical directors. This process took about 25 years.

Occupational Categories

To depict career achievements in tables 10.7–10.10, the occupations specified in the surveys were divided into four levels. Level 1 was composed of manufacturers, businessmen (mainly contractors in public works and railroad construction), and higher executives: company directors, presidents, vice-presidents, and divisional managers. Not

Table 10.6
First position held by graduates of the Ecoles d'arts et métiers, 1815–1890

Field of Work	1815–1830		1830–1890		Total	
	(n = 129)	percent	(n = 300)	percent	(n = 429)	percent
Industry and Transport						
Manufacturers and businessmen	12	9	38	13	50	12
Engineers, managers, supervisors	1	1	17	6	18	4
Draftsmen and employees	10	8	84	28	94	22
Foremen, workers, and artisans	76	59	130	43	206	48
Public and Military Services						
Technical (Ponts et Chaussées)	19	15	10	3	29	7
Employees and clerks	2	2	0	—	2	—
Teachers	3	2	2	1	5	1
Military	6	4	19	6	25	6

Sources for Tables 10.6, 10.7, 10.9, and 10.11: *Annuaire, Bulletin administratif,* and *Liste générale alphabétique et par promotions des anciens élèves des Ecoles nationales d'arts et métiers* (Paris: A & M, 1900 and 1923) published by the Société des Anciens Elèves des Ecoles d'Arts et Métiers, now called the Société des Ingénieurs . :ts et Métiers, 9 bis, avenue d'Iéna, Paris 16.

all of those listed as manufacturers in level 1 were captains of industry, for it was sometimes difficult to distinguish large from medium businessmen. Level 2 was composed of engineers, plant managers, department heads, and medium businessmen. There was some overlap between levels 1 and 2 because many graduates listed their occupation simply as engineer without specifying the actual position. These were placed in level 2, though some may have belonged in the higher level.

Level 3 was composed of works and production managers, technicians, and small businessmen. Levels 2 and 3 also overlapped, for many acquired the title of engineer when promoted to intermediate supervisory positions as shop supervisors (*chefs d'ateliers*), production managers (*chefs de fabrication*), and planning directors (*chefs de bureau*), positions placed in level 3. However, those who designated themselves simply as "engineer" appear to have been signifying a fairly high position, for it took longer to reach level 2 than level 3. So level 3 should not be seen as a subaltern category; it designates middle management. Level 4 in turn is composed of skilled personnel and lesser supervisors: foremen, machinists, fitters, and locomotive engineers, plus various employees and technicians. The four levels can thus be defined as follows: level 1, upper middle to upper; level 2, middle to upper middle; level 3, middle; and level 4, lower middle.

Occupational Attainments During the Nineteenth Century

Table 10.7 deals with graduates employed in industry, the railroads, and the public services toward the end of their careers, defined as the last position held at least 29 years later. This study of 490 graduates is based on the alumni association *annuaires*, which, beginning in 1848, provided information on the place and date of birth, the *promotion*, the occupational title, and the employer. This information was supplemented with biographical data drawn from obituaries published in the association's *Bulletin administratif*, the publications of other professional associations, and by government files in the Ministry of Industry and Commerce.

Table 10.7 indicates a high rate of success on the part of graduates, especially in industry. In all fields taken together (see column 4), nearly two fifths (39 percent) ended up in the top level, one fourth in each of levels 2 and 3 (25 percent and 24 percent, respectively), and one tenth (10 percent) in level 4. In industry one half (50 percent) reached level 1, and over one fourth (28 percent) reached level 2,

Table 10.7
Occupational attainments of 490 graduates of the Ecoles d'arts et métiers, Classes of 1820–1880

	Industry		Railroads		Public[1]		Totals	
	(n = 351)	percent	(n = 63)	percent	(n = 67)	percent	(n = 490)	percent
Level 1: Manufacturers, businessmen, higher executives	175	50	10	16	07	10	192	39
Level 2: Engineers, managers, medium businessmen	100	28	05	08	17	25	122	25
Level 3: Works and production managers, technicians, small businessmen	45	13	37	59	34	51	116	24
Level 4: Foremen, skilled workers, draftsmen, employees	31	09	11	17	09	13	51	10
Miscellaneous:[2]	—	—	—	—	—	—	09	02

[1] The public services consisted of government employment, teaching, and the military. In these fields and in the railroads, the title of engineer and the top grades of inspector belonged in level 1.
[2] The miscellaneous category consisted of *propriétaires ruraux*, artists, lawyers, and so on.

while 13 and 9 percent remained in levels 3 and 4, respectively. In other words, the great majority (78 percent) reached the two higher levels by the end of their careers, a significant achievement.

Table 10.7 covers only the nineteenth century (classes of 1820 to 1880) and was based perhaps too heavily on the alumni association rolls and its obituary notices. The membership was relatively small and was limited mainly to Paris and other industrial cities; hence the association may not have been fully representative of graduates until after 1880, when it began to grow and become more national. Therefore two further studies were undertaken to verify the findings in table 10.7. As we will see in the following sections, the results of all three studies proved to be similar.

The Poulot-Trouillet Study: The Class of 1850

Table 10.8a is a reconstitution of two studies. The first was done in 1880 by Denis Poulot on his class at Châlons (1847–1850). This was followed in 1882 by Ernest Trouillet's study on his class at Angers (1847–1850).[6] Poulot managed to trace all but 1 of his classmates. However, his information on 2 was fragmentary and 2 others died early in their careers, so the survey was reduced to 125 graduates, or 96 percent of the class. Trouillet listed 107 graduates from his class at Angers. He failed to trace 13 and 4 others had died, so the survey was based on 90 graduates (85 percent). Taken together, Poulot and Trouillet provided the occupations of 91 percent of their classmates (215/237). Of these, four fifths were traced to their last positions at least 29 years later. The remaining one fifth had died or retired before 29 years had elapsed, but they had worked long enough to establish themselves in their careers.

The results of the Poulot-Trouillet study shown in table 10.8a support the findings shown in table 10.7. The industrial column indicates that the class of 1847–1850 had slightly higher percentages reaching level 1 (55 percent versus 50 percent in table 10.7) and level 4 (13 percent versus 9 percent) and slightly lower percentages in level 2 (23 percent versus 28 percent) and level 3 (9 percent versus 13 percent). The graduates in the Poulot-Trouillet study appear to have done somewhat better in the railroads (33 percent versus 24 percent in levels 1 and 2) and somewhat worse in the public services (21 percent versus 35 percent in levels 1 and 2). The totals of all categories taken together were remarkably similar in the two tables. Thus Poulot concluded that "forty percent are the heads of industrial

Table 10.8a
The Poulot-Trouïlet Study of 1880–1882: Occupational attainments of 215 graduates of Châlons and Angers (Class of 1847–1850)[a]

	Industry		Railroads		Public and Military		Totals	
	(n = 128)	percent	(n = 33)	percent	(n = 48)	percent	(n = 215)	percent
Level 1	71	55	08	24	02	04	81	38
Level 2	29	23	03	09	08	17	40	19
Level 3	11	09	17	52	16	33	44	20
Level 4	17	13	05	15	22	46	44	20
Miscellaneous	—	—	—	—	—	—	06	03

[a] Student sample = 91 percent (215/237).

firms" who had reached their position "on the intellectual capital amassed solely at the Ecoles d'arts et métiers."

Aix-en-Provence

To round out the Châlons-Angers study, I took 185 graduates of Aix-en-Provence from the classes of 1846 to 1856 as listed in a ministerial survey in the Aix dossier in the *Archives Nationales* (*AN*).[7] This gave the positions held by graduates from the school's founding in 1843 (graduating class of 1846) down to the end of 1856. The data are presented in table 10.8b. I traced the careers of the 185 Aix graduates subsequent to 1856 in the *annuaires* and in the *Liste générale* of 1900 and combined their data (n = 185) with those from the 215 graduates of Châlons and Angers presented in table 10.8a. This makes it possible to illustrate the last positions at least 29 years or later of 400 graduates of the mid-century *promotions* of all three schools (table 10.8c).

Turning first to table 10.8b, we see that a high percentage of the 185 Aix graduates worked in the railroads: 26 percent (49/185), as opposed to 15 percent (33/215) from Châlons and Angers (table 10.8a); and they were more likely to reach level 1, whereas fewer graduates of Aix reached level 1 in industry (49 percent) than those from Châlons and Angers (55 percent). Aix was also the main supplier of naval technical officers. The navy had begun to recruit gadzarts at mid-century to oversee the mechanization of the fleet. Though regular officers looked down on their technical colleagues, calling them *bouchons gras*, the navy gradually established equivalence of ranks in order to facilitate recruitment. This can be seen in the following data, which illustrate the situation before and after the key reform of 1892:

Naval Technical Corps		Regular Naval Officers
Ranks Before 1892	Ranks After 1892	Equivalence
No Rank	Inspecteur général	Capitaine de vaisseau
2 Inspecteurs	6 Inspecteurs	Colonel
10 Mécaniciens-en-chef	20 Mécaniciens-en-chef	Chef de bataillon
70 Mécaniciens principaux, 1ère classe	100 Mécaniciens principaux, 1ère classe	Capitaine, 1ère classe
80 Mécaniciens principaux, 2e cl.	200 Mécaniciens principaux, 2e cl.	Lieutenant, 2e cl.

Table 10.8b
Occupational attainments of 185 graduates of Aix-en-Provence (Classes of 1846–1856)[a]

	Industry		Railroads		Public and Military		Totals	
	(n = 87)	percent	(n = 49)	percent	(n = 42)	percent	(n = 185)	percent
Level 1	43	49	16	33	07	17	66	36
Level 2	27	31	11	22	10	24	48	26
Level 3	07	08	13	27	14	33	34	18
Level 4	10	12	09	18	11	26	30	16
Miscellaneous	—	—	—	—	—	—	07	04

[a] Student sample = 26 percent (185/ca. 700).

In 1932 the navy finally created the rank of *ingénieur général*, equal to that of admiral. Prior to that time about 80 percent of the technical officers were graduates of the Ecoles d'arts et métiers; after that, about 95 percent.[8]

Table 10.8c summarizes the results of tables 10.8a and 8b on the 400 graduates of Châlons, Angers, and Aix. The findings closely correspond to the results of table 10.7 on the 490 graduates from 1820 to 1880. The industry category (column 1) is almost identical (50 percent in level 1 of table 10.7 versus 52 percent in table 10.8c; 28 percent versus 26 percent in level 2; 13 percent versus 8 percent in level 3; and 9 percent versus 13 percent in level 4). In both tables levels 1 and 2 combined come to 78 percent. The other categories (railroads and public and military services) and the totals correspond almost as well.

One Thousand Graduates: 1820–1920

Terry Shinn and Harry Paul have suggested that by 1900 a three-tiered hierarchy of technical schools had come into existence.[9] The Ecole polytechnique, its affiliates, Mines and Ponts et Chaussées, the Ecole centrale, the Ecole supérieure d'électricité (1894), and the Ecole supérieure d'aéronautique (1909) stood at the upper end of the hierarchy. The new engineering schools in the science faculties belonged in the middle, while the Ecoles d'arts et métiers occupied the lower end of the scale. Even the engineering diploma of 1907 made little difference, according to Shinn, for it was not widely recognized in the country and not particularly respected in industrial circles.[10] The Arts et Métiers were secondary institutions whose diploma was not even given equivalence to the *baccalauréat* until 1920 and whose graduates frequently started out as draftsmen and sometimes even as skilled workers.

This view of the schools as relatively low-level institutions training technicians rather than engineers reflects the attitudes of most French scholars and of a group of critics going back to Arago, Corbon, Le Play, and Paul Nizan. Although these critics admitted that a few exceptional individuals managed to rise to the top, they argued that these individuals had done so mainly on their own merits, while the great majority were stuck in routine middle-level supervisory positions as shop and production managers and as roundhouse masters.

It was of interest, therefore, to set up a general computer study covering both the nineteenth and twentieth centuries. The problem

Table 10.8c
Last position at least 29 years later of 400 graduates of Aix, Angers, and Châlons (Classes of 1846–1856)

	Industry		Railroads		Public and Military		Totals	
	(n = 215)	percent	(n = 82)	percent	(n = 90)	percent	(n = 400)	percent
Level 1	114	52	24	29	9	10	147	37
Level 2	56	26	14	17	18	20	88	22
Level 3	18	08	30	37	30	33	78	20
Level 4	27	13	14	17	33	37	74	19
Miscellaneous	—	—	—	—	—	—	13	03

Source for tables 10.8 a–c: *Bulletin administratif de la Société des Anciens Elènes des Ecoles d'Arts et Métiers*, 187 (June 1880), pp. 311–319, and 216 (November 1882), pp. 776–786.

Table 10.9
Last positions at least 29 years later in industry and transport of the 417 graduates of the Ecoles d'arts et métiers (Classes of 1820–1920)[a]

	Before 1881		1881–1904		1904–1920[b]		Totals
	(n = 125)	percent	(n = 131)	percent	(n = 161)	percent	(n = 417)
Level 1	35	28	42	32	77	48	154
Level 2	51	41	61	47	53	33	165
Level 3	20	16	12	09	26	16	58
Level 4	19	15	16	12	5	3	40

[a] In the same study, 63 graduates were employed in the civil service, teaching, and military: 20 (14 percent) before 1881; 17 (11 percent), 1881–1904; 20 (11 percent),1904–1920, plus 6 miscellaneous, total 480.
[b] The combined data for the periods 1881–1904 and 1904–1920 were as follows: 41 percent (119/292) in level 1; 39 percent (114/292) in level 2; 13 percent (38/292) in level 3; and 7 percent (21/292) in level 4.

was to find a means of obtaining a representative sample of the careers of all graduates over 100 years. This was made possible by the alumni association's two special editions of the *annuaire*, the *Listes générales* of 1900 and 1923. The 1923 edition contained information on all graduates since the founding of the schools, or 25,000 entries. However, 40 percent of these contained only the names, place and date of birth, and the *promotion* of the graduates. I randomly selected 1 graduate of 50 (or 1,000 names) and obtained career information on 602 of them. Of the 602 graduates for whom there were job entries, 480 had more or less complete employment histories, including the last position held at least 29 years after the beginning of their professional lives. Of these, 417 (87 percent) worked in industry and transport and 63 (13 percent) in the public and military services. The 417 are the subjects of table 10.9, while the 480 are found in table 10.10, which compares all of the tables (10.7–9) on last jobs at least 29 years later. There are in table 10.10 a total of 1,370 subjects, or 9 percent of the 15,000 graduates for whom professional data were available.

It was also necessary to find the right computer format for career biographies stretching 30 to 40 years in various fields for men working over several generations. Because job citings could vary from one to ten positions held over the course of a career, I avoided the fixed-line format of SPSS. An offshoot of SPSS called Report Generator allows for a variable-line format useful in prosopographical studies of groups over several generations. In many cases I could not find more than one or two positions held during the course of an individual career,

Table 10.10 (Summary of Tables 10.7–9)
Comparative table of last positions at least 29 years later in all fields combined of 1,370 graduates (Classes 1820–1920)[a]

	1820–1880, Table 7		1843–1856, Table 8c		1820–1920, Table 9		Totals	
	(n = 490)	percent	(n = 400)	percent	(n = 480)	percent	(n = 1370)	percent
Level 1	192	39	147	37	159	33	498	36
Level 2	122	25	88	22	187	39	397	29
Level 3	116	24	78	20	77	16	271	20
Level 4	51	10	74	19	51	11	176	13
Miscellaneous	09	02	13	03	06	01	28	2

[a] This table includes those graduates working in the public, teaching, and military sectors.

so I organized the study on the basis of years after graduation into two periods before and after 1881. I further subdivided the period since 1881 into two periods, "1881–1904" and "since 1904" (classes 1905–1920) to account for the new engineering diploma and professional advancement during this century.

Turning first to table 10.9, which contains the principal findings of the analysis of graduates from 1820 to 1920 for industry and transport (the railroads) combined, one can see that more than two thirds of those working in industry and transport before 1881 reached the top two levels (level 1, 28 percent, and level 2, 41 percent). Only a minority were in levels 3 and 4 (16 and 15 percent, respectively). The percentage reaching level 1 (28 percent) is not entirely consistent with the corresponding data in table 10.7 (1820–1880), in which industry and transport came to 45 percent in level 1 (based on an averaging of columns 1 and 2 in table 10.7: 50 percent industry and 16 percent railroads) or in table 10.8c (1846–1856), which came to 46 percent (52 percent industry and 29 percent railroads). This discrepancy suggests that the latter tables may reveal an exaggeratedly high success rate. As noted, the data in table 10.7 might be especially open to question because it includes information from obituary notices in the alumni association's *Bulletin administratif.* Although tables 10.8a–c reach very similar results to table 10.7, the *promotions* of the 1840s studied by Poulot and Trouillet may have been unusually successful, although this seems unlikely.[11]

There are two explanations for the difference in results in level 1 in table 10.9 as opposed to tables 10.7 and 10.8a–c. More than one fourth (26 percent) of the 417 graduates in table 10.9 were employed in the railways and in transport, an unusually high figure, as opposed to 17 percent of the 490 and 400 students in tables 10.7 and 10.8a–b, respectively. Gadzarts did not do as well in the highly bureaucratized railways as they did in industry. Second, in the data for table 10.9 far more graduates were listed as engineers with no specification of the exact position. I placed them in the second level, though doubtless some belonged higher. If we combine the first two levels in industry and transport, we find a close concurrence between all the tables: in table 10.7 (1820–1880) 70 percent in industry and transport reached level 1 or 2 (290/414); in table 10.8c 70 percent did so (208/297); and in table 10.9 (before 1881) 69 percent did so (86/125).

Note in comparative table 10.10, which includes the public and military services as well as industry and transport, that the findings on last positions at least 29 years later demonstrate considerable similarity from study to study. Taking all fields into account, more than one third of graduates (36 percent) finished their careers in level 1, slightly less than one third did so in level 2 (29 percent), one fifth (20 percent) in level 3, and over one tenth (13 percent) in level 4. Between 1820 and 1920 almost two thirds of 1,370 graduates surveyed in all fields reached executive, managerial, and independent ownership positions by the end of their careers.

Note also in table 10.9 that from 1881 to 1904 (column 2), graduates continued to improve their professional status: nearly one third (32 percent) reached level 1 and almost four fifths reached levels 1 and 2 combined (79 percent). Between 1904 and 1920 this figure increased to almost half (48 percent) in level 1 and over four fifths (81 percent) in levels 1 and 2 combined. Once the graduates had obtained the engineering diploma, they seldom started out as skilled workers and foremen. At the turn of this century the Ecoles nationales professionnelles replaced the Ecoles d'arts et métiers as institutions training skilled workers, foremen, and technicians. Thenceforth very few gadzarts terminated their careers in levels 3 and 4. By the 1920s the great majority were either *ingénieurs maison*, that is, promoted on the job, or *ingénieurs-diplômés*, certified engineers, and by definition they finished their careers in level 1 or 2.

The data contained in table 10.9 do not indicate how many reached the very top of the industrial hierarchy, the upper echelons of level 1. However, we can provide a more detailed analysis to describe the upper echelons of this level. From 1880 to 1920 4 percent of graduates from the Ecoles d'arts et métiers employed in industry became senior executives—presidents and company directors (*administrateurs* and *présidents-directeurs-généraux*)—and 13 percent became vice-presidents. The 1956 survey of the metallurgical and mechanical industries indicated that 15 percent of graduates reached such positions as opposed to 10 percent among graduates of all French engineering schools.[12] An alumni association survey in 1978 stated that 1.2 percent were, in that particular year, *présidents-directeurs-généraux* and 6.6 percent *directeurs d'usine*; 35.2 percent were *ingénieurs principaux et divisionnaires*, and 57 percent were *ingénieurs* (of whom 60 percent were young). It did not specify private owners, although the journal *Arts et Métiers* indicated that these had become rare. The association feared that the percentage of graduates reaching the top of the corporate hierarchy might be declining because of the rapid growth and centralization of French firms in recent years.[13] Another study done in 1977 by the national federation of engineers (FASFID) found that graduates of the Arts et Métiers earned 4 to 8 percent above the average for engineers in France.[14]

An alumni association survey in 1983 stated that about 20 percent of graduates eventually become presidents or directors of companies, 15 percent in large companies and 25 percent in small and medium firms. Another 30 to 40 percent finish their careers as vice-presidents or as divisional or regional managers. Generally, 39 percent (6,500) are employed in small and medium firms and 46 percent (7,750) in large companies having more than 1,000 employees. The remainder work in the public, military, and professional sectors. Of the 800,000 "petites et moyennes entreprises" in France, 45,000 of them account for nearly two fifths of production, one fourth of exports, and half of employment in French industry, and it is in these firms that gadzarts are particularly active.[15]

Thus we can conclude that gadzarts have held their own, and probably improved their position somewhat, since the middle of the nineteenth century, but they start higher and do not have to advance as far to reach supervisory and executive positions. During this century gadzarts have been more confined within corporate hierarchies and have had fewer chances to go into business on their own, but the persistence of small and medium companies in France has provided them with opportunities not always available in the larger firms. The attainment of the engineering diploma in 1907 and of university and *grande école* status since World War II have certainly enabled the Ecoles d'arts et métiers to consolidate their position in the face of increasing competition from other technical schools and institutes.

Other studies done since World War II continue to confirm that gadzarts do well in industry. In 1960 Nicole Delefortrie-Soubeyroux investigated the educational background of high-level management of 46 private companies (25 textile, 10 metallurgical and mechanical, 3 each electrical and food, 2 automobile, 1 petroleum, and 2 miscellaneous). She reported that of a total of 651 executives possessing higher-education diplomas, 19 percent were graduates of the Ecoles d'arts et métiers (121), 10 percent of the Ecole centrale (68), and 7 percent of the Ecole polytechnique (47). She then studied the educational background of 956 high-level managers of 12 nationalized enterprises. Of these, 25 percent (239) came from Polytechnique, 15 percent from Centrale (140), and 12.5 percent from the Arts et Métiers (120). In the public sector most gadzarts worked for the Renault automobile company, nationalized after World War II, in which they formed 71 of the 129 *cadres supérieurs* (55 percent).[16]

Remember, however, that the Ecoles d'arts et métiers produced more graduates than the other schools. For example, in the 1956 survey of the Union of Metallurgical and Mechanical Industries, gadzarts constituted 13 percent of engineers and executives working in these industries, the centraliens 5 percent, and polytechniciens 3 percent. The survey found that 46 percent of the graduates of Polytechnique, 31 percent of the graduates of Centrale, and 15 percent of the graduates of the Ecoles d'arts et métiers who were employed in these industries were senior executives.[17] The gadzarts did not enjoy the career opportunities of the polytechniciens and centraliens but were ahead of the graduates of the "petites grandes écoles" and of the universities.

Getting Ahead: The Pace of Advancement

Thus far we have discussed the first jobs of gadzarts and their last positions at least 29 years later. How long did it take them to get ahead in industry, transport, and the public services? Using the same database as for table 10.9 on the 417 graduates employed in industry and transport, I traced sequentially their professional advancement in 5-year blocks over the course of their careers, before and after 1881. Before 1881 three fourths of the 125 graduates began their careers on the bottom level (level 4). This usually meant that graduates started out as either skilled workers or draftsmen. They then became foremen and, having worked 10 to 15 years (ages 30–35), were usually promoted to various supervisory and managerial functions (level 3). After about 25 years (ages 45–50), most moved from level 3 into levels 2 and 1. By the thirtieth year (gadzarts in their early fifties), levels 1 and 2 had both passed levels 3 and 4 and by career's end had reached 28 percent in level 1 and 41 percent in level 2. This reflected a rate of advancement well beyond what the goals of the schools envisaged, but the process was not an easy one. The gadzarts had gradually worked their way up from the ranks without the benefit of a prestigious education or well-placed connections.

Among the 292 graduates of the classes 1881–1920, a higher percentage than before 1881 (41 percent versus 28 percent) reached level 1 by the end of their careers, and level 2 held about even (39 percent versus 41 percent). Taken together the percentage reaching the upper two levels came to 80 percent after 1881, as opposed to 69 percent before 1881. It is striking that the differentiation of the first two levels from the lower ones came much sooner after graduation for the classes of 1881–1920 than for those before 1881. Most graduates during the years 1881–1920 started out as draftsmen rather than as skilled workers, and they moved into engineering positions sooner. Level 2 passed both levels 3 and 4 within 12 years after graduation, and level 1 passed them 17 years after, compared to 18 years for level 2 before 1881 and 29 years for level 1 before 1881. In turn, between 1881 and 1920 level 1 passed level 2 after 30 years, when the graduates were in their early fifties (41 percent versus 39 percent).

Graduates of the classes of 1881–1920 were therefore more successful in advancing themselves than had been their fellows before 1881; a higher percentage reached the top, and they did so faster. Remember, however, that before 1881 gadzarts were of more modest origin than after. Prior to 1881 15 percent of the graduates' fathers could have been assigned to level 1 (table 10.1); since then about one fourth have belonged there. The Ecoles d'arts et métiers clearly became de facto engineering schools long before they acquired the diploma in 1907. For a generation they were in fact, if not officially, schools of engineering, and this was reflected in a relatively more prosperous clientele and in more rapid advancement to technical management posts after 1880. By the latter decades of the nineteenth century, it was no longer appropriate to describe the schools as training only foremen, technicians, and shop supervisors.

Where They Were Born and Where They Worked

Each school recruited from the surrounding departments of the region in which it was located. Châlons, and later Cluny and Lille, drew students from the north and east, Angers from the west, and Aix from the south. Before the creation of the three new schools beginning in 1900, students from the Seine and Seine-et-Oise went mainly to Angers. Though each school recruited from the surrounding region, certain industrial departments, especially in the north and northeast, were producing a higher percentage of students by the second half of the nineteenth century, notably the Seine and Seine-et-Oise (17 percent), the Nord (8 percent), the Meurthe-et-Moselle (6 percent), the Pas-de-Calais (5 percent), the Vosges (4 percent), and the Bouches-du-Rhône, Bas-Rhin, and Saône (all 3 percent). Châlons obviously recruited mainly from major industrial departments, while Angers and Aix drew their students in smaller numbers but more uniformly across the departments of their regions. Generally, then, students came from all parts of the country, and roughly half left their home region to work elsewhere in France or abroad.

Most students were of urban origin. During the second half of the nineteenth century, 15 percent came from villages of fewer than 1,000 inhabitants; 31 percent from towns of 1,000 to 5,000; 29 percent from smaller cities of 5,000 to 20,000; 15 percent from cities of 20,000 to 100,000; and 10 percent from big cities of over 100,000 (mostly Paris). At that time a population of 1,000 inhabitants was usually considered the dividing line between rural and urban, which means that 85 percent of gadzarts were of urban origin. If one sets the figure at 5,000, 54 percent were still urban in an age when most Frenchmen lived in small towns.

Most regions have always given up more young men than they have received in graduates. Before 1880, for example, the west and southwest (16 departments each) produced 18 percent of graduates and received 11 percent in return; the center (15 departments) produced 11 percent and saw only 4 percent return. The 16 departments of the southeast held even at 11 percent each way. The 14 departments of the east (12 after 1870) produced 24 percent of graduates and received 13 percent. The 9 industrialized northern departments produced 22 percent and received 23 percent in return. The big winner was the Paris region (Seine, Seine-et-Oise), which provided 11 percent of the students and saw one third live there. The remainder (3 and 5 percent) came from or worked in the Colonies or abroad.[18]

By 1890 44 percent of the alumni association members worked in six departments: the Seine and Seine-et-Oise, 31.4 percent (1,165); the Nord, 5 percent (179); the Loire, 3 percent (109); the Rhône, 2.7 percent (100); and the Bouches-du-Rhône, 2.3 percent (84).[19] The Meurthe-et-Moselle would soon bring the number to seven departments. The discovery of the Lorraine iron ore deposits in the 1890s and the rise of large metallurgical industries saw rapid growth in the number of association members residing there, from 35 in 1890 to 332 in 1909, greatly outstripping the growth in the association over these years (3,800 to 8,000).[20]

In 1975 the percentage of all graduates living in the Paris area came to 35.5, reaching a peak of 43 percent in 1977 and falling steadily thereafter to 34 percent in 1982. By 1982 the north (11 percent) and east (12 percent), the home of traditional industries, which have been in difficulty in recent years, had begun to decline in appeal to gadzarts, while the sunny, high-tech south continues to advance (the southeast, 10 percent; the Mediterranean departments,

10 percent; and the southwest, 8.4 percent). The west and center (7.5 percent and 6 percent) provide more students than they receive. Though gadzarts cluster in seven or eight industrial departments, notably the Seine, they are more likely to reside in the provinces than the average engineer belonging to the national federation of engineers (FASFID), where in 1975 more than one half (51 percent) lived in the Paris region.[21]

Fields of Work

Let us turn now from place of residence and titles held to the actual industries and fields in which gadzarts were employed over the years. Table 10.11 illustrates the number of positions held in various industries by 602 graduates of the years 1820–1920. These graduates worked for an average of five companies in 2.7 industries (602/1646) over the course of their careers, mainly in the mechanical, metallurgical, and railway industries. As one can see in table 10.11, industry and transport accounted for 87 percent of the positions held by the graduates of these classes. The percentage employed in the mechanical and metallurgical industries grew from 32 percent before 1881 to 45 percent between 1881 and 1920, with each field about equally represented. On the other hand, the percentage working in the railroads and other forms of transport declined from 26 percent before 1881 to 10 percent between 1881 and 1920. This resulted from the end of the great railway boom of the middle decades of the nineteenth century and the bureaucratization of the companies, which closed avenues of promotion.[22] After 1904 only 5 percent worked in the railways, and since World War II only 4 percent have done so. Interest in the public and military services has declined in similar fashion.

According to the alumni association (1983), an estimated 26,000 gadzarts are alive today. Of these, 24,575 can be traced, of whom 16,925 (69 percent) are active professionally in France. The number working in the mechanical industries has risen over the years to one third (34 percent or 5,700/16,925), as opposed to 24 percent between 1881 and 1920 (table 10.11), while metallurgy has declined from 21 percent (1881–1920) to less than one tenth (8 percent, or 1,350) in 1983. Increasingly popular today are electronics, computers, and information services (5 percent, or 850), business and financial services (11.5 percent, or 1,950), and teaching (7 percent, or 1,200). One third of graduates are scattered over several other industries and

Table 10.11
Number of positions held in various industries by 602 graduates of Ecoles d'arts et métiers, Classes 1820–1920[a]

	Before 1881		1881–1920		Totals	
	(n = 508)	percent	(n = 1138)	percent	(n = 1646)	percent
Industry and Transport						
Bank, business, insurance	3	—	32	3	35	2
Building, public works	50	10	63	5	113	7
Electricity, gas, water	20	4	102	9	122	7
Mechanical, auto-aviation	75	15	270	24	345	21
Metallurgy, mining	84	17	243	21	327	20
Petroleum, chemicals, tire and rubber	0	—	40	4	40	2
Railroads, transport	132	26	109	10	241	15
Textiles	26	5	51	4	77	5
Misc. Industry[b]	51	10	79	7	130	8
Total	441	87	989	87	1430	87
Public						
Government	25	5	8	—	33	2
Military	25	5	94	8	119	7
Teaching	17	3	47	4	64	4
Total	67	13	149	13	216	13

[a] The gadzarts worked in 2.7 (602/1646) industries and in an average of five companies over the course of their careers.
[b] Miscellaneous included food industries, forest products, glass and so on.

fields. Of those who were not professionally active, 850 graduates (3 percent) live outside of France in 64 states and territories; 6,000 (24 percent of 24,575) are retired, and 800 (3 percent) are continuing their education or doing military service.[23]

We have shown in this chapter how the graduates of the Ecoles d'arts et métiers have reached much higher positions, especially in industry, than the schools' modest official goals and reputation as working-class institutions suggested. Ninety percent reached supervisory posts by the end of their careers (levels 1 through 3), and approximately one half reached upper-middle to upper-level managerial positions (level 1). They often did so in small and medium companies as well as in the larger corporations. Over the course of their careers they passed through all echelons of the corporate hierarchy: skilled workers, draftsmen, technicians, and production engineers. In the process they contributed substantially to industrial development and played a crucial role as intermediaries between the bosses in the front office and the workers on the factory floor. Though some rose quite high in the larger companies, the key to their success lay in their role as managers, directors, and owners of small- and medium-sized firms manufacturing various specialized machine tools, mechanical appliances, engines, motors, and parts. In the following chapter we will discuss gadzarts at work: how they built their careers and how some managed to build small companies into flourishing concerns.

Gadzarts in Industry

On vit de bonne soupe et pas de beau langage
Traditional saying of gadzarts

The Ecoles d'arts et métiers were originally trade schools that evolved over the nineteenth century into schools of mechanical engineering. Before the award of the engineering credential in 1907, however, the gadzarts were obliged to advance themselves on the job, changing positions several times over the course of their careers and trying their hand variously at civil, mechanical, and metallurgical engineering. But essentially they were production engineers concerned with the manufacturing process and with the design, construction, installation, and maintenance of heavy machinery and power plants. In 1899 Louis Bouquet, director of the technical education division in the Ministry of Industry and Commerce, said of them: "They are no longer employed as workers but rather as production supervisors and plant managers. Many of them become directors of important companies, creating brilliant positions for themselves in the various branches of industry at home and abroad."[1]

Gadzarts were proud of their contributions to industry, their role in the building of a national transport system, the spanning of rivers and valleys by bridges and viaducts made of structural iron and steel, the mechanization of shipping and port facilities, and the big public works projects. Later in the century they pointed to the harnassing of electrical energy and internal combustion and the rise of the electrical, automotive, and aviation industries. In the following pages we will outline their contributions during the past two centuries in these and other fields. We must concentrate, in Nizan's words, on a few "heads of grain" that rose higher than the others, but it is nevertheless useful to trace the pattern of their careers, for almost all started out at modest levels and worked their way up in industry and transport.

Machine Construction

Despite France's lead in engineering in the late eighteenth century, the state engineering schools contributed little to the mechanization of industry, which was mainly the work of self-taught artisans and mechanics, British and French. Machine building was an appendage of other industries—mainly coal mining, metallurgy, and textiles— that required mechanical equipment and machines. In metallurgy the Le Creusot works, followed by the Fourchambault Company in the Nivernais, Denain in the Nord, Decazeville in the Aveyron, and de Wendel in Lorraine were the first to introduce English-style puddling furnaces and rolling mills. Yet by the middle of the nineteenth century France was still smelting more than half her iron in small charcoal furnaces scattered in wooded areas around the country. This limited the demand for machinery and equipment and retarded the growth of the machine-construction industry into the 1830s. Wood, wind, and water continued to be the main sources of energy, and steam power advanced only slowly.

The earliest and most important manufacturers of machines during the eighteenth century were Jacques-Constantin and Auguste-Charles Périer, who manufactured steam engines and pumps and who introduced the Watt engine in France during the 1770s. During the revolutionary and Napoleonic period, with ties to England cut off, industry made little progress. In 1810 there were only 200 Watt-type stationary engines in France. After 1814 the Risler brothers, Dixon, Jean Meyer (Châlons, 1832), and André Koechlin manufactured machines in Alsace for the textile industry. Hallette had a shop in Arras; the gadzarts Michel Cazalis and Charles Cordier, backed by La Rochefoucauld, opened one at Saint-Quentin; and Manby and Wilson established a firm at Charenton that introduced English techniques. Gadzarts Pierre Saulnier established himself in Paris, as did Calla, Cavé, Decoster, and Edwards.[2]

Most of the original manufacturers in the machine-construction industry were not formally educated. As James Edmonson has shown, only 6 of 88 members of the Union des Constructeurs in the 1840s possessed diplomas from engineering or trade schools.[3] First-generation industrialists were suspicious of school-trained workers and generally preferred to train their mechanics and foremen on the job. Much of the actual work of production planning and execution was done by jobbers, that is, by foremen who subcontracted work from the entrepreneur (*marchandage*). The owner furnished tools, equipment, and materials, while the foreman organized production and hired and paid the workers, assuring that each job was completed on a schedule set by the owner's works manager.[4] In this sense the early factory was really an assortment of individual shops. Self-taught foremen feared the competition of school-trained mechanics like the gadzarts and were reluctant to hire them.

However, the growth of the railway industry in the 1840s and 1850s stimulated both metallurgy and machine construction. The number of steam engines in France grew from 10 in 1810 to 2,900 in 1840, 5,400 in 1846, and 7,800 in 1852. These were located mainly in five industrial departments, in the north (Nord and Seine-Inférieure), the northeast (Haut-Rhin), the southeast (Loire), and the Seine. Of the 50 machine-building companies operating in the mid-1840s, 7 were capable of producing locomotives: Koechlin in Mulhouse, Schneider in Le Creusot, de Wendel in Lorraine, Hallette in Arras, Cail in Paris, and two new companies: La Ciotat near Marseille, financed mainly by the Talabot brothers to build locomotives and ships, and the machine-construction company at Batignolles, financed by the Railroad Company of the Nord and directed by Ernest Gouin.[5]

The railways brought profound changes in industrial organization. The companies demanded precision machines and standardized, interchangeable parts, which encouraged the development of special-purpose machines for planing, slotting, turning, and boring machine tools. This caused manufacturers to reassess their attitudes toward technical education. Workers and foremen on the factory floor seldom had the knowledge of mechanics and thermodynamics necessary to study the resistance of materials to heat and stress, which was of crucial importance in the manufacture of locomotive boilers. Nor could they easily learn the latest techniques in mechanical drawing at a time when the measured drawing was beginning to replace the model as the method of analyzing machines and their components. Hence the draftsman and the production engineer began to come into their own during the 1830s and 1840s.

Since 1816 the Petite Ecole of the Conservatoire des arts et métiers of Paris had been training draftsmen in Monge's method of mechanical drawing. The director, V. Le Blanc, joined in 1829 by the gadzarts Jacques-Eugène Armengaud (Châlons, 1825), who succeeded him in 1835, not only trained draftsmen for the mechanical industries but established the Conservatoire as the center of the diffusion of industrial technology. Le Blanc founded the *Recueil des machines, instruments et appareils*, published from 1819 to 1852 and edited by Armengaud from 1829. Armengaud founded and published the *Publication industrielle des machines, outils et appareils* from 1841 to 1890, the leading journal of industrial technology in France during the period.[6]

In addition to his role as professor of mechanical drawing at the Conservatoire, Armengaud was a partner in a machine-construction company with his father-in-law, Nicolas-Guillaume Cartier, and was also the director, with his brother Charles (Châlons, 1828), of an important firm of consulting engineers specializing in industrial patents and railway technology. In his various enterprises Armengaud demonstrated the importance of technology to industry and the useful role of mechanical drawing in production planning and execution. He was, for example, an early advocate of steam cutoff (*détente*) in the design of locomotives, which saved fuel and made the engine more efficient. He campaigned to reform France's outdated patent laws and to introduce uniform screw threads and the standarization of parts.[7] Because of his access to the wealth of materials at the Conservatoire and his experience as a machine builder and patent consultant, he possessed, in the words of Edmonson, "an unrivaled knowledge of technological precedents"; his *Publication Industrielle* had "no serious rival as a record of current manufacturing techniques in the engineering industry."[8] In 1856, on the recommendation of leading machine builders Bourdon, Cavé, Chapelle, and Farcot, he was inducted into the Legion of Honor, the second gadzarts in industry to receive that honor (Jules Houel was the first in 1849). Of modest origin, the son of a noncommissioned officer in the Napoleonic army, he demonstrated in his career how the graduate of a modest school for workers could win rapid promotion in the emerging profession of industrial engineering.[9]

The first firm to hire gadzarts in large numbers was the machine-building company of Charles Derosne and Jean François Cail. Cail, the son of a coppersmith and himself a former boilersmith, had little formal education. In 1834 he joined the chemist Derosne in establishing a machine shop in Paris specializing mainly in the manufacture of mechanical equipment for the sugar beet industry. Thanks to the investments of the Lebaudy brothers, with whom both Derosne and Cail were linked by marriage, new factories were opened in Belgium in 1838, Denain in 1844, and in Valenciennes, Douai, and Amsterdam by 1847.[10]

In 1844 Derosne died and Cail became the sole proprietor of the firm. That same year he obtained a contract from the Railroad Company of the Nord to produce locomotives. He hired Jules-César Houel, a graduate of the Ecole d'Arts et Métiers of Angers in 1832, as chief engineer in charge of planning and supervising the establishment of a locomotive factory in Paris, which involved the installation and coordination of more than 200 machine tools. Houel, one of the first technical school graduates hired by Cail, began his career in the mid-1830s as a fitter before being promoted to foreman and shop supervisor. As chief engineer in charge of the Paris works, he supervised the manufacture of hundreds of locomotives (between 1856 and 1866, the company produced 800 locomotives).[11]

Houel introduced important changes in industrial management in the Cail company. He organized a planning office (*Bureau d'études*) staffed by engineers and draftsmen, which replaced the foreman on the factory floor as the focus of production planning and control. Henceforth, the foreman's job was to oversee the execution of plans and drawings sent down from the planning office. As James Edmonson noted:

"Houel used drawings to complete the design of a machine so that its manufacture could be broken up into a myriad of separate tasks assigned to foremen and their crews of mechanics. By so doing, he transferred greater responsibility to his engineers. They had to work out the definitive form a machine would take before the draftsmen took over. The design function was thereby preempted by the engineers, effectively removing it from the shop floor.[12]

Houel's "managerial revolution" used mechanical drawings to organize and allocate work. The planning office became the control center in large machine-building enterprises. This system, combined with the scientific organization of machine tools in order to maximize production, changed the hierarchy of skills in the machine-building industry and enabled Cail to efficiently manufacture locomotives on a large scale.

Houel's system allowed for greater specialization of labor and speed of execution; however, this made it difficult to train skilled personnel on the job and undermined apprenticeship. As mechanical precision replaced manual dexterity, the fitter (*ajusteur*) began to disappear and the setup man (*monteur*) became more important. The monteur had to be adept at reading blueprints. Aside from the Petite Ecole of the Conservatoire des arts et métiers, which had relatively few students, the Ecoles d'arts et métiers were the only schools training students who could read blueprints with facility, who were excellent draftsmen, and who also understood the principles of mechanics and thermodynamics.[13]

The gadzarts' ability to conceptualize machines and their components made them invaluable in the production process and explains why manufacturers began to hire and promote them in large numbers in the 1840s. J. F. Cail, the first major employer of gadzarts, hired about 15 per year and averaged 100 on the payroll, including 6 or 7 department heads, at any one time. He sent his son and successor, Alfred, to Châlons in 1856. He encouraged Houel to hire the best graduates of the Ecoles d'arts et métiers, some of whom became chief engineers and directors of Cail's various enterprises and later important industrialists in their own right, notably Edmond Beaudet (Châlons, 1847), Ferdinand Caillet (Angers, 1835), F. E. Collignon (Châlons, 1845), Antoine Montigny (Angers, 1849), and Félix Moreaux (Châlons, 1843).[14] When Cail transformed his company into an anonymous society in the early 1860s, the Société française de construction mécanique, Houel, followed by Beaudet, Caillet, Montigny, and Moreaux, left Cail and formed a partnership with Parent and Shaken of Lille in the organization of an important new machine-construction company called Fives-Lille. Fives had plants in Lille and Givors (Rhône) and more than 10,000 employees in the early part of the twentieth century.[15]

Though young gadzarts commonly began their careers at Cail or Fives, which they referred to as their "écoles d'application," many then moved on to other companies or established themselves as proprietors (usually in partnership with other gadzarts) of small and medium machine and metal shops. The *Bulletin mensuel* of the alumni association was full of advertisements for the sale of machine shops and small firms ranging in price from 20,000 to 200,000 francs.[16] The ease with which gadzarts moved back and forth between small and large companies (the average graduate changed jobs five times during his career) illustrates the nature of the "dual economy" of the nineteenth century. While large companies moved toward centralization and standardization in fields that lent themselves to mass production, a network of smaller workshops and firms survived by manufacturing specialized products and by subcontracting arrangements with larger companies. The small shops employed master craftsmen operating general-purpose machines, and they sometimes prospered by inventing new products and introducing new industrial processes.[17] Such small companies persisted in all Western countries, despite the growth of big industry, but the small firm appears to have flourished particularly in France. France's consumers were more attached to regional products, her craftsmen were more closely associated with artisanal techniques, and her labor force was more reluctant to leave the countryside for the city.[18] This "dual economy" suited the gadzarts, who were willing to move between small and large firms, from region to region, and between machine-construction, metallurgical, engineering, and railway companies.

Whether they worked in large or small companies, gadzarts enjoyed increasing success in industry from the 1840s on. Ernest Gouin, a polytechnicien and the director of the Compagnie de Batignolles, a major producer of railway locomotives and equipment, told the commission investigating technical education in 1864 that graduates of the Ecoles d'arts et métiers were sought after by manufacturers "because of the scarcity of men with a thorough knowledge of machines who are also good artists and draftsmen. . . . You don't get this kind of man from the Ecole polytechnique and the Ecole centrale; these gentlemen expect to rise quickly to the top, while the students from Châlons and Angers are more patient."[19] In its concluding report to the 1864 investigation, the commission observed that gadzarts had played a key role in the development of French heavy industry: "Without their participation our big metallurgical, railroad and mechanical industries could never have developed independently of foreign technology; today England and other countries come to France in search of engineers and technicians."[20] All this of course facilitated their advancement to engineering posts. As Arthur Morin wrote in 1860, "We can cite a large number of high-ranking industrialists who come from the Ecoles d'arts et métiers. Promoted originally by their bosses, who were struck by their practical abilities, they have frequently risen to the head of big companies."[21]

Heavy Engineering and Public Works

The first graduate of the Ecoles d'arts et métiers to gain a national reputation in civil engineering was Nicolas Cadiat (Châlons, 1820). The son of a metalsmith, inventor of a number of steam engines and motors, and the designer of big metal bridges, he had become by 1829 a director of the metallurgical works at Fourchambault, where he collaborated with Polonceau in the design and construction of the Pont du Carrousel in Paris. After working for a time for the Koechlin Company in Alsace, he joined with Dietrich in organizing a firm specializing in the manufacture of locomotives. When that company was absorbed by Cail in 1842, he became chief engineer of the metal works in Decazeville in the Aveyron. In 1848 he joined with a fellow gadzarts, Oudry (Châlons, 1832, Ecole polytechnique, 1835), former chief engineer of the Ponts et Chaussées, in opening a consulting firm; they designed the Pont d'Arcole of Paris and the massive swing bridge of Brest, major works of the time. Cadiat died prematurely in 1857, just as the company was gaining full momentum.[22]

Félix Moreaux joined Cadiat and Oudry in 1854, participating in the construction of the Pont d'Arcole. When Cail took over the firm, Moreaux became director of planning. He moved on to Fives-Lille a few years later, where he specialized in bridge construction. He designed bridges on the Cher River at Montluçon and on the Allier at Moulins, in which he developed a system of self-reinforcing treliced girders, dispensing with vertical girders, which made the bridges more graceful in appearance. Moreaux was one of the first engineers in Europe to sink pylons by means of forced air. He subsequently designed and built the Pont de l'Europe in Paris, the Zwolle Bridge in Holland (474 meters, with central crossbeams of 75 meters and a swing section of 41 meters), and the Plaisance Bridge on the Po River in Italy (580 meters).[23]

Among other gadzarts successful in the public works industries, Louis Barret (Aix, 1844) supervised the mechanization of the port of Marseille. Vincent Dauzats (Angers, 1856) assisted Ferdinand De Lesseps in building the Suez Canal, in which 18 gadzarts participated. Léon Chagnaud (Châlons, 1881) supervised the building of the canal and tunnel of the Rove linking Marseille with the Rhône, one of the great engineering feats of the time. He planned and executed the extension of the Paris metro under the Seine and the deep stations of the Cité and Saint-Michel, as well as the three lines running through the Place de l'Opéra.[24] His colleague, Léon Ballot (Angers, 1885), built several of the first big dams in France, notably on the Cure and the Dordogne. Simon Boussiron (Aix, 1888) pioneered in the development of reinforced and prehardened concrete. He designed the elegant Pont de Montauban over the Tarn (1911) and the Pont de La Roche-Guyon over the Seine (1935), utilizing new methods and materials.[25]

The Railways

The railway act of 1842 laid out the lines radiating from Paris to the main provincial cities, but the 30 or so companies in the field were undercapitalized and were set back by the depression of the 1840s. Costs were high, and skilled personnel had to be imported from Britain. By mid-century France had built only 3,000 kilometers of railroads.[26] The Second Empire encouraged railway construction and the concentration of industry; the period saw the rise of six great railroad companies: Nord, Est, Ouest, Midi, PLM, and Paris-Orléans. The machine-construction companies were producing 500 locomotives per year by 1854 and were exporting them by the 1860s.[27]

More than any other group, the gadzarts built and ran the French railroads, but they seldom did so as directors and chief executives of railway companies, who were usually polytechniciens and centraliens. They were instead the engineering directors of private companies such as Cail and Fives (Houel), which built locomotives and parts for the railways, private entrepreneurs who constructed the sections, the railway beds, bridges, and viaducts under contract (Cadiat, Oudry, Moreaux), or consulting engineers in the technology of railway building (Jacques and Charles Armengaud).

Other graduates worked for the railroad companies, but the more ambitious tended to leave after a few years in favor of industry. The railroads, indirectly controlled by the state, were more bureaucratic in nature.[28] Advancement to top positions as inspectors and engineers of the line was slow and depended on seniority. In 1879 there were 369 graduates of the Arts et Métiers working for the PLM, mostly in the elite traction division, but only about 8 percent at any one time held top positions as engineers, inspectors, and divisional heads. Most had to be content with posts as depot masters, shop supervisors, and foremen (44 percent), while the rest were mainly locomotive engineers, mechanics, and draftsmen.[29]

The most successful of the *gadzarts cheminots* of the nineteenth century was probably Louis Martin (Châlons, 1837). Starting out as a worker, he became a *conducteur des ponts et chaussées* in 1844 at the age of 23 and was assigned to the group planning the Paris-Strasbourg line. In 1849 he joined the Railway Company of the East as a section head and was soon promoted to district chief. When the Emperor Napoleon III expressed the wish to visit the military installation at Mourmelon in 1857, Martin studied and built in just 65 days the 27-kilometer line going from Châlons to Mourmelon and crossing the Marne River and Canal. He was rewarded with the position of *Ingénieur principal de la voie* and the Legion of Honor at the age of 36.

During the War of 1870 Martin helped construct siege works and later helped rebuild the city of Paris. During the Prussian siege he designed and built 34 flour mills. From 1874 to 1896, when he retired, he was chief engineer of the Vincennes line. He was president of the alumni association of the Arts et Métiers from 1866 to 1875, and in 1884 he was the first gadzarts to be elected president of the Société des ingénieurs civils.[30]

Iron and Steel

In the field of metallurgy the Ecoles d'arts et métiers produced several of France's leaders, notably Hippolyte Petin (Châlons, 1828) and Camille Cavallier (Châlons, 1871). Born in Amiens in 1813, Petin was the son of a small merchant. After graduating from Châlons, he worked as a draftsman and then as a fitter in Lyon for 50 *sous* per day. In 1837 he met fellow metalsmith Jean Gaudet and established a small forge at Rive-de-Gier in the Loire Department. On the advice of an engineer of the Ponts et Chaussées, they produced iron parts for the Givors Canal then being built toward Saint-Chamond. Having succeeded in this and other ventures, they acquired enough capital to invest in the new steam hammer just invented by Bourdon (chez Schneider) and Nasmyth in England, which enabled them to produce large iron pieces for machine-construction companies, the railroads, and the navy.[31]

In 1850 Petin and Gaudet opened a steel mill at Saint-Chamond (Loire) and in 1854 they merged with Jackson and Company of Assailly (Loire). The new company was called the Compagnie des Hauts Fourneaux, Forges et Aciéries de la Marine et des Chemins de Fer. Petin traveled to England to inspect the new Bessemer converter and soon installed one in the mills at Assailly and Saint-Chamond. In the Crimean and Franco-Prussian wars the company became a major arms producer, especially of heavy artillery and big warships. By 1865 the company employed 5,300 workers in seven plants and produced 50,000 tons of iron and steel. The company's working capital rose from 500 francs in 1837 to 37 million in the mid-1860s. Drawing on mineral deposits from large holdings in Corsica and Sardinia, the company manufactured steel girders, iron rails, sheet metal, big cannons, chassepot rifles, heavy armor, and bayonets.

In 1874 Petin and Gaudet retired, having created one of the largest metallurgical companies in France. Petin's sons, Charles and Jules, and son-in-law, Emile Roux, remained active in the firm, though as directors rather than owners, for the company had become a large anonymous corporation under the direction of Adrien de Montgolfier, a *centralien*, who remained in the post for 35 years until his death in 1908. Around the turn of the century the firm expanded into Lorraine and became the *Société de la Marine et d'Homécourt*. It joined the *Sidelor* group (*Union Sidérurgique de Lorraine*) after World War II.[32]

Among other successful nineteenth-century metallurgists, Lucien Arbel (Aix, 1843) and Claude and Barthélemy Brunon (Aix, 1851 and 1852) were owners of major metallurgical and arms companies at Rive-de-Gier.[33] Antoine and Agamemnon Imbert (Châlons, 1835, Aix, 1850) owned an important company at Imphy (Nièvre) that produced girders, boilers, and rails, mainly for the PLM Railway in the southeast. Earlier in his career, Antoine Imbert had worked for Petin and had supervised the installation of the Bessemer converter in the Assailly mill.[34]

Among other major metallurgical companies of the nineteenth century, the Schneider family at Creusot and the de Wendels in Lorraine frequently hired graduates. One hundred sixty-five gadzarts were working for Schneider in 1900.[35] Schneider's factory school for young workers at Creusot sent its best students to the Arts et Métiers, most of whom returned to the company. The firm of François de Wendel was also a major employer (50 in 1934 in the Meurthe-et-Moselle). Albert Bosment (1864–1960), a graduate of Châlons in 1880, followed his father, Anthime Bosment (Châlons, 1854), as director of the main plant at Hayange; he then became director of the *Forges de Joeuf* and finally director-general of mines and factories. He and his father spanned the period 1858–1938 as directors at Wendel (1858–1912 and 1883–1938). Albert Bosment's daughter married another gadzarts, René Paschal (Lille, 1902), who succeeded his father-in-law as director-general and who presided over the formation of various big steel mergers following World War II, *Sollac* and *Sidelor*.[36]

Perhaps the most outstanding industrialist of all graduates was Camille Cavallier (Châlons, 1871), for many years director of Pont-à-Mousson, one of France's leading metallurgical companies. Cavallier was the son of a modest forest service employee in a small town in Lorraine (Meurthe-et-Moselle) who had moved there from the Hérault in southern France. He attended a local primary school and the municipal college of Pont-à-Mousson. After graduating from Châlons he worked in a small foundry in Pont-à-Mousson directed by another gadzarts, Xavier Rogé (Châlons, 1850). Rogé had found it difficult to procure sufficient coke from dwindling supplies in local forests, and, especially after the loss of the Lorraine deposits to Germany in 1871, of iron ore. After visits to England he decided to specialize in the manufacture of cast iron piping and tubing for carrying water, sewage, gas, and other liquids, and for light metal construction in structures such as *les Halles*. Rogé's decision to move in this direction made the fortune of the company. After 1900, under Cavallier's administration, Pont-à-Mousson became a worldwide supplier of metal pipes and tubing.

Rogé and Cavallier also played an important role in the discovery of vast iron ore deposits in French Lorraine. Everyone had assumed that the Germans had successfully taken all the iron ore of the Lorraine region in the territorial settlement of 1871. But Cavallier had a hunch that the deposits of the Briey region continued westward into French territory and persuaded his colleagues to explore at Auboué near Homécourt. Iron was indeed found, but it was of poor quality. Cavallier insisted on continuing the probes, and in 1883 the famous "couche grise" was discovered, four meters in thickness, of excellent iron ore. The company then began the long and expensive process of extracting coal and iron from the marshy soil of the area.[37]

In 1900 Cavallier became sole director and inaugurated a period of rapid modernization and expansion. By 1913 the export of tubing, pipelines, and castings accounted for 60 percent of French exports in the field (68,000 of 117,690 tons).[38] With the German invasion of 1914, the mill at Pont-à-Mousson was destroyed by bombardment and the factory and mine at Auboué occupied. Cavallier transformed the foundry at Foug into a war plant near the front and established others at Saint-Etienne du-Rouvray near Rouen and at Belleville (Meurthe-et-Moselle) and Sens (Yonne). He was planning a factory at Toulouse when the war ended.[39]

With the termination of hostilities, Cavallier headed a major company with eight factories and several mines. He transformed it into the *Société anonyme des hauts fourneaux et fonderies de Pont-à-Mousson*. Despite the major war damage and cost of reconstruction, in 1922 the company still accounted for 60 percent of France's export of pipelines, tubing, and cast iron (64,000 tons).[40] Cavallier's ideas on reconstruction were similar to those of planners after the Second World War: he wanted to set up a national planning commission to allocate raw materials and establish priorities, a process requiring close cooperation between the state and the private sector. He envisioned this as including a program of extensive social legislation and a nationwide campaign against alcoholism, poor housing, and a declining birthrate.[41]

As director of the company for 52 years, Cavallier experienced only one strike, in 1905, as labor unrest swept across Lorraine. In 1906 he established a permanent board of arbitration with representation of workers and employees. He also introduced health insurance, retirement pensions, seniority bonuses, recreation centers for workers, and housing for retired workers and employees. His stated purpose was "to build a moral as well as material association based on a sense of belonging and of common purpose and responsibility." When asked the secret of his success, he said, "You pay careful attention to detail, and you never rest." When offered various honors and decorations, he refused "to be turned into a flower pot." His son and son-in-law were polytechniciens, but his board of directors was about equally divided between them, centraliens, and gadzarts. His younger brother, Henri (Châlons, 1884), his sons, and his grandsons continued to direct the company. It exists today as PAMCO and is still a major employer of gadzarts.[42]

Electricity

Another field in which gadzarts demonstrated considerable originality in the application of science to industry was electricity. Hippolyte Fontaine (Châlons, 1848), one of 13 children of a carpenter in Dijon, began his career as a carpenter and then as a draftsman in Lyon. He was gradually promoted to mechanical engineer and became interested in electricity on his own. After the War of 1870 he joined the company of Zénobe Gramme, the Belgian inventor of the dynamo. The dynamo had no immediate industrial applications until Fontaine demonstrated its reversibility in 1873. Soon after, at the Vienna Exposition, Fontaine demonstrated electric power distribution by powering a motor with the current provided by a generator located 1,800 meters away.[43]

Camille Faure (Aix, 1857) and his partner Nicolas Raffard (Angers, 1841) developed a number of industrial applications from new discoveries in electricity during the latter decades of the nineteenth century. Faure introduced a storage cell of sufficient capacity to be used industrially and developed the first electric furnace (1883) capable of producing aluminum, a metal just beginning to be manufactured. He and Raffard built battery-powered cars used in factories and mines. They also manufactured electrical generators employing the discoveries made by Fontaine.[44]

Joachim Estrade (Aix, 1873) exploited the invention of the reversible dynamo and the generator by transporting energy on a large scale. In 1900 his company, the *Société Méridionale de Transport de Force*, distributed 20,000 volts of electrical energy over a distance of 160 kilometers from Carcassonne to Narbonne. He was among the first to succeed in dividing up a large force transmitted at high tension into smaller quantities at low tension, through the use of substations. As a result, the company could sell electrical power to individual and corporate customers scattered over 100 cities and towns and reaching 100,000 inhabitants along the way.[45]

Charles A. Keller (Angers, 1890) pioneered in the field of electro-metallurgy at the beginning of this century. He developed a process for converting scrap metal into cast iron by means of his electrical furnaces. During the Great War, when France had lost much of its metallurgical industry and its iron ore and coal reserves to the Germans, Keller's plant at Livet in the Isère was an important source of iron and steel for the armaments industry. After the war he was instrumental in the development of the hydroelectric industry in the Alpine region of southeastern France.[46]

Automobiles

During the early years of the automobile industry, the Ecoles d'arts et métiers produced a number of outstanding inventors and manufacturers. Charles Trépardoux (Châlons, 1868) was the partner of De Dion and Bouton in a company manufacturing mechanized tricycles and autocars. Emile Delahaye (Angers, 1859) began to produce automobiles in 1896 and was the first to use the water pump to help cool automobile engines. Henri Brasier (Châlons, 1880) founded his company in 1897, a forerunner of Simca, and introduced four-wheel brakes in 1908. His cars, of advanced design, won several international racing meets in the early years of this century.[47]

Louis Delage (Angers, 1890), the son of a minor railroad employee, reached prominence as the designer of racing cars that won several grands prix. He founded his own company in 1905 and by 1926 had 2,500 employees in his mechanized factory at Courbevoie in a Paris suburb. His luxury cars ceased to sell during the Depression, and he went bankrupt at about the same time as André Citroën, two of France's most innovative car makers.[48] Finally, Louis Coatalen (Cluny, 1895) became a director and partner in the Sunbeam Company in England during the early years of this century. During the war he played a major role in building aviation and truck engines for the British and American armies. He retained his interest in France and was a partner in the Anglo-French company Sunbeam-Talbot-Darracq.[49]

With the concentration of the automobile industry since the Second World War, gadzarts have been most successful as technical and production directors of Citroën-Peugeot and Renault. Louis Renault disliked the fancy *grandes écoles* and hired and promoted many gadzarts. For example, Alphonse Grillot (Châlons, 1902) worked for Renault for more than 50 years, becoming director of production and later director-general from 1945 to 1965. He helped reorganize the company after nationalization in 1945 and was associated with the mass production of the *Dauphine* in the 1950s. Another director, Pierre Bézier (Paris, 1927), supervised the mechanization of the Renault assembly lines during the 1960s.[50]

In the field of petroleum engineering Eugène Houdry (Châlons, 1908) was a prolific innovator. After the First World War he worked to develop a fuel suitable for the smooth running of internal combustion engines. After several years research he invented a new process of refining crude petroleum, which he called catalytic cracking. But Houdry could not obtain necessary financial backing in France for his process, and in 1931 he moved to the United States where, backed by the Standard Oil Company, he founded the Houdry Processing Company in which he implemented the system of catalytic cracking. The system involved three simple processes: polymerization, cracking, and refining; it resulted in a more highly refined and higher-octane fuel at a lower price.

The arms race of the 1930s and the Second World War vastly increased the demand for high-octane fuel, especially aviation fuel. The Houdry Processing Company sold licences to 67 firms and held more than 600 patents on everything from catalytic cracking to the manufacture of butadene, a chemical essential to the making of synthetic rubber. In his later years Houdry became concerned with the problem of air pollution and patented a number of devices to limit emissions. He was so vehemently anti-Vichy that the government withdrew his French citizenship, and he ended up becoming an American.[51]

Armaments

During World War I the gadzarts actively participated in the task of converting French industry to mass production for a war of attrition. In the face of devastating human and material losses in 1914, the success of French industry in quickly adapting to a war economy may be considered the first "economic miracle" of the twentieth century, and though it is difficult to define the gadzarts' role in this collective achievement, one may point especially to the fields of artillery and aviation, in which they made outstanding contributions.

Victor Champigneul (Angers, 1874) had built his father's small foundry into a large metallurgical firm. During the war he developed highly mechanized machine presses for mass producing artillery shells, machines that were used in every major war plant in France. His own company produced 225,000 bullets and projectiles and 60,000 cannon shells daily. In 1915 the minister of Armaments appointed Champigneul to coordinate the production of the big arms companies.[52]

Along with Champigneul, René Guillery (Châlons, 1883) was a key innovator; his precision machines, used especially in the production of ball bearings, accelerated the arms output. Jules Robin (Châlons, 1867) spent his life doing arms research at the Ecole militaire de prototechnie at Bourges (1876–1919), inventing 60 different explosive shells, many used during the war. C. G. Bessières (Aix, 1898) also invented explosive shells, notably the V.B. grenade, of which 300,000 per day were manufactured during the war.[53]

In the field of marine technology, Jean Fieux (Cluny, 1902) did pioneering work in the gyroscope for the Schneider Company at Le Creusot and for the French navy, which led to rapid advances in the stabilization of warships, crucial while firing big guns. Among his many other inventions, Fieux developed a catapult/brake system during the interwar period that made it possible to launch airplanes in the confined spaces of the aircraft carrier.[54]

Aviation

The gadzarts' most spectacular contributions came in aviation. On the eve of World War I this field was still in its heroic period. There were no wind tunnels to test materials and flying conditions, and there was no mass production; each plane had to be individually crafted. It was difficult to produce a propeller, used so successfully under the sea, that could gain sufficient traction in the air, and it was equally difficult to develop a motor that was light and yet powerful enough to thrust the machine into the air. Overheating was a problem that obviously had to be overcome.

In 1908 Lucien Chauvière (Angers, 1891) patented a laminated wood propeller that powered a Blériot airplane right across the English Channel on July of 1909. During the war he produced 100,000 propellers, one fourth of French production, and later introduced metal and aluminum models. Louis Verdet (Aix, 1885) invented a rotary engine that had great thrust, was light, and was easy to cool.[55]

During the war Albert Mary (Lille, 1905) designed the Nieuport fighter plane in 1915, which attained the unheard of speed of 150 km/h, a third faster than anything else in its class. He then designed a smaller, even faster model called the *Bébé* (13 meters), which the English also adopted. He followed that in 1917 with the Nieuport 28, used by the Americans. Mass production of the series was supervised by Jules Legros (Châlons, 1900).[56]

The fastest, safest, most maneuverable French fighter plane manufactured during the war was the Spad, a sturdy, simple machine that was easy to produce in quantity (13,000 during the war). It was designed by Louis Béchereau (Angers, 1896), assisted by a team of gadzarts. The pilot Guynemer, France's ace, favored the plane. On the occasion of Béchereau's induction into the Legion of Honor, Guynemer wrote to him: "You have given our country air supremacy, and you will play a major role in our victory."[57]

The best observation plane was the Bréguet 14 (5,000 produced during the war), which was designed by Marcel Vullièrme (Cluny, 1904) and proved its worth in the Battle of Verdun. The first bombers were designed by Emile Coquelin (Lille, 1907), technical director of the Farman Company, and by Mongermon (Angers, 1906), technical director for Avions Voisin. The first torpedo hydroplane was designed by Antoine Odier (Aix, 1901), who with Bessières founded the Ecole supérieure des travaux aéronautiques in 1931.[58] By the end of the war the French aircraft industry had produced 100,000 airplanes. Marius Vernisse (Cluny, 1913), director of France's Air Arsenal before World War II, wrote: "Before, during and after the Great War, of forty-two factories working for national defense and for aviation, forty-two directors, heads of planning staffs, and production managers were gadzarts." The six most important manufacturers of airplanes, airplane motors, and dirigibles all had graduates of the Arts et Métiers as technical directors:[59]

Béchereau, Avions Spad
Vullièrme, Avions Bréguet
Mary, Avions Nieuport
Mongermon, Avions Voisin
Canton, Moteurs Salmson
Cormon, Etablissements Zodiac (dirigibles)

After the war the government's interest in aviation declined. Most efforts were private, directed toward cargo and transport, scientific research, or adventure and sport. Mary designed a new series of Nieuports. André Herbemont (Châlons, 1909) produced the excellent Spad-Herbemont in 1918 and, as technical director of Spad-Blériot, the first retractible landing gear in France in 1930.[60] Vullièrme designed the Bréguet 19 (2,200 produced), which flew in stages across the North Atlantic in 1924, and the Bréguet-Bidon, which later flew from Paris to New York. Moine (Angers, 1910) designed several models of the Laté, immortalized by Saint-Exupéry. The models produced by the *Société Potez* of this period were the work of Louis Coroller (Angers, 1909) and a team including more than 50 gadzarts.[61]

René Couzinet (Angers, 1921) built his first plane at the age of 22. He then designed and built a trimotor, the *Arc-en-Ciel*, which Mermoz piloted across the southern Atlantic in 1933. Despite its superior handling, safety, and economy, which made it an ideal long-distance mail carrier, the government ignored it, and Couzinet was still trying to get it into production when war broke out in 1939.[62]

Until 1936 the French government paid little attention to the concept of air warfare. But the occupation of the Rhineland and the Spanish Civil War convinced some officers and politicians that the Maginot Line would provide inadequate protection against the Blitzkrieg. In 1937 the Air Ministry was created, including an Air Arsenal organized under the direction of Marius Vernisse (Cluny, 1913). In 1939 the armed services were coordinated under the ministry of Raoul Dautry, as they had been under Albert Thomas in 1916. The Morane MS 406 and Dewoitine D520 were good fighter planes designed under the direction of Gautier (Lille, 1912) and Vautier (Aix, 1916), but political and bureaucratic delays prevented full-scale production until 1938–1939. The Bloch 700, designed by Herbemont, was tested at a speedy 550 km/h but never got beyond the testing stage.[63]

Since World War II gadzarts have continued to be active in the resurge of French aircraft production that began in the 1950s. They were involved in planning and production in all major firms. Alfred Asselot (Aix, 1919) and André Vautier (Aix, 1916) became directors at Sud-Aviation and collaborated on the Caravelle and Concorde.[64] By 1966 about 800 graduates were working in all branches of aviation—public, private, and military. In 1981 this had risen to 1,000 of about 15,000 engineers working in the aviation industry. About 40 percent are employed in technical sectors, 30 percent in production and control, 15 percent in administration, and 15 percent in commercial and sales departments.[65]

Inventors

Among inventors Nicolas Raffard (Angers, 1841) held more than 100 patents on machine tools, locomotive prototypes, and electrical appliances and motors. The Conservatoire des arts et métiers has 15 of his motors in its collection. He is best remembered for his design of a system of electric tramways, a high-speed electric locomotive, a dynamometer, a two-piece dynamo, elastic coupling assemblages, and stop valves and plugs. At the Exposition of 1889 he won a gold medal and was introduced by the president of the Republic as "the foremost mechanical engineer in France." [66]

Léandre Mégy (Aix, 1851), another successful inventor, won 13 gold medals at various world fairs. Best known for his classic railroad brake, he held numerous patents on equipment for the elevation and extraction of heavy materials and big hoisting machinery, including various cranes and winches powered by steam and electricity. He developed a pile driver capable of driving a pile 25 meters, automatic firing devices for various weapons, and an electrically powered swing bridge that he successfully demonstrated in the World's Fair of 1889. He also invented a variety of steam and electric motors, two-way brakes, connecting gears, dynamometers, and other devices. Decorated in Italy and Spain, one of France's greatest inventors, he died in 1910 having never been honored in his own country.[67]

The saddest story of France's neglect of her own inventors is that of Guislain Goubet (Angers, 1853), inventor of the submarine. The son of a gadzarts, Désiré Goubet (Châlons, 1828), he followed in his father's footsteps as a railway inspector. Leading a comfortable middle-class life, the younger Goubet became fascinated by the challenge of undersea navigation. He submitted a project to the Ministry of the Navy and was encouraged by the minister to continue his research. At his own expense he then designed and built the first workable submarine, the "Goubet One," which he demonstrated at Cherbourg in 1891 before naval ministry officials. Although the two-man craft passed all mechanical tests and proved floatable, habitable, and submersible, the navy brass rejected it on questions of detail. Goubet then redesigned his submarine, again at his own expense. Borrowing money, soliciting gifts from patriotic citizens, he worked for eight years and in 1900 presented the "Goubet Two" to the navy at Toulon. The craft dove 20 meters and returned to port having successfully passed all tests, yet the naval bureaucracy, jealous of any invention not its own, again rejected it for administrative reasons. Overwhelmed by the blow and heavily in debt, Goubet died three years later in a charity home. Ironically, that very year the minister of the Navy, Camille Pelletan, formally recognized his achievement as inventor of the submarine and honored him post-humously.[68] In this case, as in so many others during the half century after 1890, the Germans and other rivals benefited most from French inventive genius.

Among the many gadzarts who established small- and medium-sized firms, two were particularly inventive and innovative. Edouard Granger (Châlons, 1822), the son of an indigent worker, applied his knowledge of chemical metallurgy to substitute bronze, copper, tin, and inexpensive alloys for precious metals in order to cheaply repro-duce masterworks. His costume jewelry and metal ornaments, known as *articles de Paris*, introduced a new industry in France. Granger also manufactured various inexpensive metal products for the theater (armor, swords, jewelry, stage decorations, props, and sets), which had hitherto been made of wood and cardboard.[69] Another middle-sized firm that prospered was the *Forges d'Hennebont*, founded by Emile and Henri Trottier (Angers, 1828) at Angers, the only metal-lurgical firm in the area at the time. They were succeeded by Emile's son Jules (Angers, 1857). Although the company remained small and familial, it was efficient and intelligently specialized, producing various tin products, and has survived until recent times.[70]

Politics

Graduates of the Arts et Métiers have seldom been inclined toward politics. During the nineteenth century most of those elected deputy or senator were successful industrialists of the right center politically who had started out as local mayors. Under the July Monarchy,

Henri Flaud (Angers, 1830), a successful manufacturer of fire pumps, was elected from Paris. During the Second Republic the machinist Jeandeau served as a socialist in 1848 but was not reelected in 1852.[71] Few were elected to the *Corps législatif* during the Second Empire, but the Third Republic saw more activity. A number of industrialists from the southeast sat in the National Assembly and Senate, including metallurgists Lucien Arbel (Aix, 1843), Barthélemy Brunon (Aix, 1852), and Agamemnon Imbert (Aix, 1850) from the Loire Department. Simon Plissonnier (Aix, 1864), a manufacturer of agricultural and industrial machinery, was a deputy for many years from the Isère, and Alexis Savary (Angers, 1867), elected senator in 1894, was the largest producer of agricultural machinery in Brittany.[72] From the Nord, Désiré Linard (Châlons, 1856), Alfred Macherez (Châlons, 1858), and H. P. Sirot-Mallez (Châlons, 1851) were owners of major sugar-refining companies, and Jules Deregnaucourt (Châlons, 1835) was a successful producer of textiles at Roubaix.[73]

Later on, Gaston Goudron (Angers, 1899, and the Ecole supérieure d'électricité) moved from a successful position in the electrical industry to the Armaments Ministry under Albert Thomas in 1916, where he served as head of armaments production and then as director of technical services in the Ministry of Industry and Commerce under Etienne Clémentel until 1920. He sat in the Chamber of Deputies from 1928 to 1936 and was twice named secretary of state, first in the Steeg cabinet of 1930–1931 and then in Herriot's third cabinet in 1932.[74] He was one of the few gadzarts to reach cabinet rank. Arthur Groussier (Angers, 1878) was offered a cabinet post but, as a good socialist, turned it down. Elected from Paris in 1893, he served as a deputy from the inception of the socialist party over a period spanning 30 years and later became vice-president of the Chamber of Deputies. Since World War II few graduates have been elected to Parliament.[75]

Enumerating a few score "stars" among the many graduates of the Ecoles d'arts et métiers over almost two centuries provides only a partial picture of their professional lives. Nevertheless, these examples, when taken together with the collective career portrait of the previous chapter and the testimony of many industrialists and observers, illustrate the successes that gadzarts have enjoyed in industry. The best years were probably the middle decades of the nineteenth century, when the Ecoles d'arts et métiers had the pick of the best young technical minds coming up from the primary system and from the modern programs in the secondary schools. Once the Third Republic began to multiply professional schools serving the common people, the Arts et Métiers lost their virtual monopoly and increasingly had to compete with other institutions for technical talent.

Although the Ecoles d'arts et métiers did not produce as many captains of industry as did Centrale and Polytechnique, they graduated a surprising number of successful men: for example, in machine construction Armengaud, Houel, Fontaine, and Fieux; in the automobile industry Brassier, Delahaye, and Delage; and in metallurgy Arbel, Petin, and Cavallier. The schools also produced many capable and energetic engineers who figured among France's most ingenious inventors. In any case direct comparisons between the Arts et Métiers and the *grandes écoles techniques* are misleading, for most gadzarts had to rise from humble origins and move up through specialized technical posts that provided relatively few openings to the top. Most centraliens and polytechniciens came from prominent families, and the centraliens were frequently the sons of successful businessmen and manufacturers, which gave them a considerable advantage. More important, if we take French successes in the automobile and aviation industries and the spectacular increases in production during the First World War as examples, we can conclude that the gadzarts and centraliens, and sometimes also the polytechniciens, were capable of great achievements when they put aside their competition for position and power and concentrated on the challenges of production.

12

Conclusion

The expert men, technologists, engineers, or whatever name may best suit them, make up the indispensable General Staff of the industrial system; and without their immediate and unremitting guidance and correction the industrial system will not work. . . . The material welfare of the community is unreservedly bound up with the due working of this industrial system, and therefore with its unreserved control by engineers, who alone are competent to manage it.

Thorstein Veblen

Social scientists (among them Bourdieu, Crozier, Granick, Passeron, and Suleiman) have argued that French education, in its emphasis on general culture and elite formation, neglected technical studies and provided little opportunity for social mobility.[1] In their view the graduates of technical and vocational schools lacked at once the technical skills and "cultural capital" to advance beyond lesser supervisory positions and therefore had little or no impact on industrial technology. In addition, some economic and scientific historians (Ben-David, Gilpin, Locke) have argued that the elitist and outdated programs of French education contributed to the decline of French science, technology, and economic productivity, as France fell further and further behind Germany and the United States during the second half of the nineteenth and early twentieth centuries.[2]

Few of these authors closely examined France's technical schools at either the higher or intermediate level. However, Frederick Artz, René Taton, and especially Terry Shinn and John Weiss have made valuable contributions to our understanding of the history of higher technical schools, the social origins of their students, and the evolution of the engineering profession in France.[3] J. P. Guinot and Antoine Léon have concentrated on apprenticeship and on popular professional instruction and their defects.[4] The present work has been the first to examine the intermediate technical schools, which have trained skilled workers, industrial technicians, and engineers in France for almost two centuries. Because these schools operated largely outside the mainstream of French public education, they have been overlooked by historians and observers and their impact underestimated.

We have shown that by the turn of the twentieth century, France had developed a solid system of intermediate technical schools that supplied manufacturers with the skilled technicians and managers they required and that provided the common man with opportunities for social advancement. In so doing we have traced the formation of a new socioprofessional group of middle-level technical managers in industry. Most of these were industrial technicians and engineers who were trained mainly in a set of intermediate technical schools under the Ministry of Industry and Commerce; others were educated in professional schools under various ministries or in private, mainly Catholic, schools.

The impact of the technical schools was particularly great during the years from 1880 to 1930 and since 1950. The turning point, however, came during the 1840s with the rise of the railway industry. The manufacture of locomotives and heavy machines imposed demanding technical requirements on industry, which shifted the focus of production planning and control from the foremen on the factory floor to the engineers and draftsmen in the planning office. The gadzarts Jules Houel, who directed the installation of J. F. Cail's big locomotive works in Paris, led the way in introducing new management techniques. As the mechanical industries increasingly sought to hire industrial engineers, technicians, and draftsmen during the middle decades of the nineteenth century, the demand for gadzarts and for graduates of other technical schools rose accordingly. By the second half of the century, most company directors, managers, and even owners in the mechanical industries were technically educated, of whom about 20 percent were gadzarts (chapters 10 and 11).[5]

By the 1860s a number of industrial associations and alumni groups, gathered around the Ministry of Industry and Commerce, advocated the separate organization of technical and professional schools under Commerce, called the *Université du Travail*. Although they had little success under the Second Empire, they gradually formed a technical education lobby under the Third Republic, which played an active role in the struggle during the 1890s between the Ministry of Industry and Commerce and the Ministry of Public Instruction over the higher primary professional schools. The striking parliamentary victories of Commerce in 1892 and 1900 gave it a small but complete system of schools ranging from the higher primary professional schools to the Ecoles d'arts et métiers at the intermediate level and the Ecole centrale in higher education, in short, something resembling the *Université du Travail*.

The rivalry between the two ministries set back the development of technical education somewhat but was also a sign of growing interest in the subject among educators. That the Ministry of Industry and Commerce could win major parliamentary battles with the much larger Ministry of Public Instruction and obtain substantially increased funding for its growing school system indicated increasing support in France for intermediate technical training. Despite its setbacks in 1892 and 1900, the Ministry of Public Instruction continued to expand and professionalize its higher primary schools and encourage the creation of engineering institutes in the science faculties of the universities.

The excellent performance of French industry during the First World War in gearing the war-ravaged country to fight a long war of attrition against the Germans suggests that the French were not deficient in technical personnel. After the setbacks of the interwar and war years, the growth of intermediate technical education in France since the Second World War has been impressive, though not without its problems. A mass system of intermediate and higher technical and vocational schools has emerged with 2 million students, who may choose from technical and professional lycées and collèges, two-year technical community colleges (Instituts universitaires de technologie), numerous teacher-training and vocational institutes, and various other programs and facilities for continuing education. As we have seen, this system is built on a foundation extending back to the early years of the Third Republic. Indeed, the expansion of technical and engineering education under the Third Republic, especially during the generation prior to the Great War, must rank as one of the Republic's greatest educational achievements.[6]

Our study of the Ecoles d'arts et métiers has been particularly useful in assessing the contributions of technical schools to industry and the opportunities they provided for professional advancement. Of all the intermediate technical schools during the past century, the Ecoles d'arts et métiers have been the most successful in maintaining their identity, improving their programs and diplomas, and providing occupational opportunities for their graduates. The three, and later six, schools have been the major suppliers of mechanical engineers and managers to French industry, transport, and the government technical services. Hence they have provided us with the opportunity for an in-depth study of a set of technical schools over five generations of industrial development.

Our study in chapter 10 of the origins and careers of more than 2,000 graduates between 1820 and 1920 has provided significant data on the process of social and professional mobility through technical education and industry. Despite the official definition of the Arts et Métiers before 1907 as schools training skilled workers and foremen, only a small minority (11 percent) of graduates during the nineteenth century finished their careers as foremen, skilled workers, and draftsmen. The great majority of graduates employed in industry reached supervisory and managerial posts. By the end of their careers, half of those employed in industry had become owners, directors, presidents, and vice-presidents of companies, and regional, divisional, and plant managers. During the twentieth century an average of 20 percent have become directors, presidents, and vice-presidents, 15 percent of them in large companies and 25 percent in small and medium enterprises; most of the remainder have become regional, divisional, and plant managers. The engineering diploma of 1907 and the promotion to *grande école* status since World War II have helped compensate for declining opportunities to start a business on one's own and have enabled gadzarts to compete more effectively in industrial bureaucracies with the graduates of the many new engineering schools that were established from 1890 to 1914 and since 1950.

In addition to their collective attainments, individual gadzarts have made outstanding contributions to the development of French industry. In metallurgy Hippolyte Petin founded Aciéries de la Marine et d'Homécourt; Xavier Rogé and Camille Cavallier established Pont-à-Mousson, one of France's most successful companies; and Lucien Arbel organized Arbel Frères, a major metallurgical and arms firm. In machine construction Jules Houel supervised the introduction of locomotive manufacturing *chez* Cail and in the process revolutionized management techniques in France. Houel and other gadzarts co-founded the Fives-Lille Machine Construction Company. In consulting and patent engineering, and in the diffusion of technology, the firm of Armengaud Frères was one of the earliest and most important.

Since the middle of the nineteenth century, gadzarts have provided most of France's naval technical officers. They were the technical directors of all the aviation companies producing airplanes in World War I. In automotive engineering they, along with the centraliens, have been the chief pioneers in a field in which the French have always excelled. Gadzarts have been particularly prominent among technical employees and engineers at the Renault Motor Works over the years.

Despite the monopolies of the graduates of the Ecole polytechnique and the Ecole des ponts et chaussées in public works, gadzarts have designed and supervised the construction of a great many of France's most important heavy engineering projects, especially of bridges, dams, subways, canals, and modernized port facilities. Among their many inventions, graduates introduced variable steam cutoff in locomotive engines and demonstrated the reversibility of the dynamo. They pioneered in the technology of underwater navigation, catalytic cracking, and the marine gyroscope, and they introduced thousands of patented parts, machines, and processes. So we can conclude that the Ecoles d'arts et métiers have contributed greatly to the development of French industrial technology during every stage of industrialization in France while providing an excellent means of social and professional advancement to their students (chapter 11).

The gadzarts' professional attainments were made possible because of the greater possibility in private industry for advancement for young men of modest origin than in the nationalized sectors, the railways, and the public and military services. Although the top positions in industry have tended to go to the graduates of the elite technical schools, to the centraliens and polytechniciens, engineers from other schools did have access to executive positions. Many companies in the automobile, machine, and metallurgical industries (notably Cail, Fives-Lille, Pont-à-Mousson, and Renault) have traditionally preferred to hire engineers from the Ecoles d'arts et métiers and other schools having a practical orientation.

The gadzarts' success can be attributed to several causes: the integration of intellectual and manual instruction in the school programs; the excellent recruitment (for many years the schools had a quasi monopoly over the best technical minds coming up from the working classes and sectors of the petty bourgeoisie); the *internat*, which fostered group solidarity in opposition to the Strass; the contacts of the alumni association; and the energy and resourcefulness of graduates in working their way up through the ranks of industry. However, we must also ask what role their family background and earlier education played in the formation of their attitudes and in their professional attainments.

By the time of the July Monarchy, the students were mainly the sons of skilled craftsmen in the machine and metal trades who were comparatively well paid and much sought after in industry. When they reached the Ecoles d'arts et métiers, they had received a solid primary education and were themselves craftsmen who had already done their apprenticeship. Later in the century they came to the schools as technicians trained mainly in the Ecoles nationales professionnelles and in specialized higher primary schools.

Thus the students were not ordinary workers but came rather from the petite bourgeoisie industrielle, that is, from an elite of skilled workers and small producers who were in a privileged position to benefit from the mechanization of industry and who saw the advantage of educating their sons in the applied sciences. The students were not opposed to the schools' academic goals, but they had no clear idea of their professional prospects in the new middle-level managerial class that was emerging with the centralization and bureaucratization of industry. Unfortunately, because the Strass treated them as delinquents rather than as future collaborators, and the professors kept their distance, it was difficult for young gadzarts to find role models who could guide them in the transition from the artisanal trades into industry. The schools concentrated on technical training and neglected the human and social issues attendant upon industrialization.

The administration, itself uncertain as to the values, attitudes, and skills to teach, squandered many opportunities. The Duke de La Rochefoucauld sought to define a new program combining theory and practice, small classes, work in common, and a cooperative approach to learning on the part of faculty, staff, and students, but this was rendered impossible by the system of *casernement*, which the duke was partly responsible for introducing. As a former military officer (the original school at Liancourt had been designed for the sons of soldiers of his regiment) and as an advocate of the new techniques of confinement in hospitals and prisons, he believed it possible to use the boarding school to transform the most unpromising human material into new industrial men. La Rochefoucauld embodied the qualities of aristocratic paternalism, bourgeois liberalism, and French militarism in a not unattractive way, and he was always popular among the students. However, the authoritarian and manipulative nature of the *internat*, which he organized between 1806 and

1824 at Angers and Châlons, lent itself to abuse and, over the long run, proved to be a failure.

The coming of the July Monarchy and the reform of 1832 provided an opportunity to reestablish the schools on a solid foundation. The schools ceased to recruit the sons of soldiers and petty civil servants and began to successfully appeal to those of skilled craftsmen and small producers. J. A. Vincent, director at Châlons and later inspector general of the schools, oversaw their adaptation to the steam age. Charles Dauban at Angers governed humanely and successfully from 1832 to 1848. Yet government officials were more influenced by a few weeks of trouble in the Angers school in 1848 than by 16 years of relative peace and harmony. By the second half of the century, the Strass had abandoned all pretense of leadership in favor of repression. The decision-making process was transferred upward to the far off Ministry of Industry and Commerce, creating an ethical vacuum in the schools that was filled unofficially by the student peer group and by the alumni association, founded in 1846. The peer group defined norms of behavior and the rules and rituals of clandestine self-government, while the association assumed responsibility for defining professional roles and defending the *positions acquises* of graduates.

This unstable equilibrium, characterized by frequent breakdowns of morale and discipline, led to serious criticisms of the Ecoles d'arts et métiers. Arago, Corbon, and Le Play predicted that the uprooting of young workers from home and workshop, forcing them to learn abstractions unrelated to the real world of production, would only produce bad managers and inferior human beings. More recently, Paul Nizan stressed the theme of class betrayal and dehumanization, while Michel Bouillé underlined the schools' conservative teachings as the main reason for the transformation of students into company men.[7] The alumni association has also been criticized for its role in appropriating student customs and channeling the gadzarts' spirit of solidarity in resistance to oppression into an outgoing corporate image.[8] In one way or another, therefore, all the critics stated that the schools successfully destroyed the working-class culture of the students, forming them into compliant middle-level managers. From the capitalist perspective, the disciplinary regime achieved its purpose of breaking in and remolding the young men.

At this point the ideas of Michel Foucault in *Discipline and Punish* (1975) are of interest, especially his view of *clôture* as central to behavior manipulation in the Western world since the Enlightenment.[9] Along with Nietzsche, Foucault associated the Western idea of knowledge with the will to power rather than with the Platonic quest for understanding. In achieving the large-scale rational organization of men and things, the "disciplinary society" of the modern Western world developed highly sophisticated instruments of political and ideological control. These instruments took the form of institutions run by professionals. Foucault borrowed Bentham's theme of the Synopticon, the model prison of the nineteenth century, as a symbol of a society in which penal methods of control were extended into other institutions. The spread of "technologies of coercion," which Foucault calls the "carceral archipelago," saw the isolation of subjects "for their own good" under the power of "experts" (doctors, psychiatrists, educators, lawyers, and bureaucrats). Because of the persuasiveness, the apparent service orientation, and the elevated credentials of these experts, the essentially coercive nature of the modern institutions that they control has been disguised by their apparent rationality and naturalness.

The Ecoles d'arts et métiers illustrate Foucault's theory of incarceration as a means of disciplining and transforming oppressed or "dangerous" groups by a combination of "regimentation and intimidation" and "individualization and homogenization." In addition to the more obviously repressive aspects of education, the presentation and acquisition of knowledge are, according to Foucault, political acts that train individuals by reproducing authorized knowledge and forms of research, by constant observation and classification, and by encouraging individual competition and specialization.

Foucault referred to the reformatory at Metray, established in 1840, as the best example of an institution combining elements of prison, school, and workshop, which transformed delinquents into docile subjects who did not even stir during the revolutionary events of 1848. The Ecoles d'arts et métiers, founded a generation earlier and adding the elements of barracks and cloister to those of penal institution, school, and workshop, perhaps better exemplified Foucault's theory than did Metray, except that the gadzarts were far from docile. When Cluny and Lille were established at the end of the century, the spirit of resistance spread quickly to them, as it did also to the Ecoles nationales professionnelles at Armentières, Vierzon, and Voiron. Indeed, by then many of the students of the Arts et Métiers were already veterans of three years in the *internat* of the Ecoles professionnelles, which helps to explain the constant warfare between students and Strass during this decade.

The first question that arises from Foucault's analysis is the extent to which the harsh carceral regime of the nineteenth century influenced the behavior of graduates. At first glance it appears that the "hard" and "soft" forms of conditioning under the old and new regimes (before and after the reforms of 1900) worked equally well in preparing gadzarts to accept industrial discipline. Yet one may ask whether the harsh *ancien règlement* really succeeded in molding the young men along the desired lines. In the 120 songs analyzed in this book (chapters 8 and 9) and in the speeches and writings of graduates, one finds nothing but hostility toward the Strass, especially to the principals and *censeurs*; hence it is unlikely that the Strass could have had much normative influence on the gadzarts, except in the most negative sense of forcing them into a delinquent community of opposition. In the final analysis, the gadzarts did not so much defeat the carceral regime as bring it to a standstill.

But perhaps the impact of three years in the *internat* was more subtle. Did the gadzarts show any signs of having internalized the propaganda to which they were subjected, and did their values after leaving school substantially differ from what they had been before they arrived? Given the atmosphere of conflict in the schools, it is surprising that a disaffected minority did not emerge among graduates espousing some form of sociopolitical protest. Despite their grumblings about bureaucrats and polytechniciens interfering with the production process and blocking careers open to their talents, gadzarts have seldom joined socialist parties or other protest groups. Though always Republicans, they have been generally right of center in politics and functionalist in outlook, believing in progress, efficiency, and technical education as a means of getting ahead.

Politically, gadzarts have been Republican individualists rather than socialists. Alain, with his emphasis on individual work and the concrete world of production, was more representative of them than Jean Jaurès. However, during the interwar period their loyalty to the Republic was severely tried by political instability and by the nationalizations and prolabor policy of the Popular Front governments. They have generally been content with the Fifth Republic, which promoted the Ecoles d'arts et métiers to *grande école* status and liberally funded technical education. Some, however, attached to the old ways, regret the disappearance of the technical education division and the loss of contact with the working classes (chapter 7).

Like the centraliens studied by John Weiss (chapter 2), the gadzarts have been typical engineers, much more interested in good design and technical efficiency than in abstract ideas or revolutionary theories. They believed that the capitalist system would work much more efficiently if technologists rather than bureaucrats and politicians were at the helm. As students they took what they needed from the scientific and technical offerings of the schools, but their values seem to have been mainly produced by their home environment. Because so many were the sons of independent craftsmen and small producers, they doubtless believed in private property and free enterprise before arriving at the schools. The righteous Strass, the suspicious guards, and the harsh and joyless atmosphere of the *internat* drove them to organize a student underground, a process that occurred in many French boarding schools during the century, but the gadzarts—part workers and part soldiers—were not quite like other students. Their particular style of resistance was related to their crafts background, for in many ways their ritualistic behavior, secret codes, arcane language, and cult of production recalled the guild activities of their forebearers (chapter 8).

The *ancien règlement* was probably never necessary and must be attributed to the obsessive concern for order and fear of revolution of the nineteenth-century bourgeoisie. The *nouveau règlement* was introduced at the turn of the century by Millerand, a former socialist, after several years of lobbying by the alumni association. It was only a partial improvement. Although the students henceforth enjoyed more freedom, they continued to be endlessly observed, ranked, examined, classified, and categorized in a manner characteristic of Foucault's homogenization-individualization. They were still not taught the language skills and the knowledge of business and accounting that would have facilitated their advancement, and their association with manual work and production engineering continued to limit their competitiveness in the industrial hierarchy.

The students were not fooled; they correctly perceived that liberalization represented another, more subtle form of control:

Et dans mes bons jours,
Faisant patte de velours,
Je vous accorderai un peu de liberté.
Oui, par la douceur,
Je veux gagner vos coeurs;
Je serai sûr alors, sur vous, de pouvoir régner.[10]

The new regime was nevertheless necessary because the old disciplinary system had broken down, and for all its limitations, liberalization provided a much healthier atmosphere for young people to work and study in compared to the nightmare of the *ancien règlement.* One discovers in the student songs of the nineteenth century repeated themes of loneliness, melancholy, and death, especially in the frequent theme of the monastery-school as sepulcher (le cloître silencieux, le muet fantôme, le monastère séditieux, les prisons austères, les cloîtres maudits, la morne promenade, le couvent aux froids couloirs, les vieux tombeaux). This suggests that the school experience appeared in the students' subconscious mind as a form of death, involving the loss of home, country, and workplace, of one's freedom and personal identity. In the isolation of the school and later the workplace, gadzarts were a caste of outsiders who had to go through a series of ritual trials (including symbolic death and rebirth) to prove their worth as managers and men.

Around the turn of the century the alumni association sought to break the deadlock created by the administration and students, but it faced a number of problems in doing so. As the embodiment of the gadzarts' spirit, the association was obliged to always defend the students against the Strass, which undermined public relations, especially in high places. Because of the rise of the engineering institutes in the science faculties, the appearance of several new Catholic technical schools, and the increasing *pantouflage* of polytechniciens around the turn of the century, gadzarts faced increasing competition for management positions in industry. If they were to compete for executive jobs, gadzarts had to improve their public relations, and the schools had to upgrade their programs and diplomas while projecting an image of order and propriety within. It was essential to rid the students of the stigma of working-class disloyalty, which became further implanted in the public mind each time the army was sent in to quell a riot in one of the schools.

The association could not advocate the elimination of the *internat*—the source of all common experiences—so it argued during the 1890s for the end of *clôture*, distinguishing between the system of harsh discipline and the boarding school regime itself. If *clôture* had to be eliminated as the source of violence in the schools, the *internat*, properly directed, was indispensable to the transformation of a delinquent student subculture into an esprit de corps opening out into business and industry. Once this took place, the "trads" would lose their clandestinity and, in the process of codification, much of their mystery, evolving eventually into the harmless antics and pranks normally associated with student life. In other words, the softer *nouveau règlement* proved successful because it seemed natural and reasonable.

The "Gadzarts' Spirit" of fraternity and solidarity has been embodied in the alumni association, in "the long chain" (in their language) that dates to 1846, and indeed to earlier unsuccessful attempts to win governmental approval for their society. Despite some conflicts between young and old and between employers and employees, graduates have joined the association in ever increasing numbers. And they have remained loyal to it because it defended their interests, gave them status, provided them with a sense of personal worth, and perpetuated the comradeship and solidarity of the school years, which they saw as the most significant human experience of their lives.

The Ecoles d'arts et métiers have come a long way during the past two centuries, rising from schools of arts and trades to engineering colleges to *grandes écoles techniques*. They have been, and to some extent still are, the elite schools of the upper working class and small industrial bourgeoisie seeking to get ahead in industry. From the professional viewpoint, the schools have successfully trained competent engineers and managers for industry, transport, and the government technical services for almost two centuries. From the human perspective they have been less successful. Rather narrowly vocational, isolated from the outside world, and run on a penal-military model for their first century, they were separated from the mainstream of public education, the major centers of scientific and industrial research, and even industry itself. Although the barracks regime fostered an extraordinary solidarity that carried over into the careers of graduates and into the life of the alumni association, it also created a delinquent community of students distrustful of the outside world and of authority and impeded their later socialization in industry and private life. The energy, inventiveness, and solidarity of the gadzarts probably resulted as much from their background in the crafts and small industry as from the school experience.

In theory the gadzarts should have looked back in anger on "la vie morose" in the isolated and somber setting of the monastery-schools and on the smoking factories where they had spent so much of their lives. They had been caught first between the unsmiling Strass and the implacable "elders" and later between the workers who resented their schooling and the disdainful bosses reluctant to admit them to their ranks. While they have shown signs of having an inferiority complex as a result of their uncertain status, the gadzarts have also been rather satisfied with their professional attainments. They have attributed these to their education, to their fraternity in the face of oppression, to hard work and enterprise, and to the industrial system that provided them the opportunity to get ahead and that they have so faithfully served over the years. As to their inner doubts and private lives, one can only conjecture, for they have seldom talked about themselves publicly. Corbon and Le Play correctly argued that the school experience destroyed the craftsman so that a new breed of man, neither worker nor boss but a salaried manager, could rise from his ashes. As La Rochefoucauld had foreseen, the gadzarts became new industrial men, the technologists and engineers who have managed the industrial machine and made it work.

Notes

Chapter 1

1. Paul Gerbod, *La condition universitaire en France au XIXe siècle* (Paris: Presses Universitaires de France, 1965). For general works on French education, see Félix Ponteil, *Histoire de l'enseignement en France, les grandes étapes 1789–1965* (Paris: Sirey, 1966); Antoine Prost, *Histoire de l'enseignement en France 1800–1967* (Paris: Armand Colin, 1968); and Joseph Moody, *French Education Since Napoleon* (New York: Syracuse University Press, 1978).

2. On the genesis of the French engineering profession and the social origins of students, see especially Terry Shinn, *Savoir scientifique et pouvoir social, l'Ecole Polytechnique, 1794–1914* (Paris: Presses de la fondation nationale des sciences politiques, 1980), pp. 196–216, and John Weiss, *The Making of Technological Man: The Social Origins of French Engineering Education* (Cambridge, Mass.: The MIT Press, 1982). For the early works on French science and technical education, see Frederick Artz, *The Development of Technical Education in France 1500–1850* (Cambridge, Mass.: The MIT Press, 1966), and René Taton, *Enseignement et diffusion des sciences en France au XVIIIe siècle* (Paris: Herman, 1964).

3. Charles Dupin, *Avantages sociaux d'un enseignement public appliqué à l'industrie* (Paris: Bachelier, 1824), and Arthur Morin and Henri Tresca, *De l'organisation de l'enseignement industriel et de l'enseignement professionnel en France* (Paris, 1862).

4. Michel Crozier, *The Bureaucratic Phenomenon* (Chicago: University of Chicago Press, 1964), pp. 145–208; Pierre Bourdieu and J. C. Passeron, *La Reproduction: éléments pour une théorie du système d'enseignement* (Paris: Les Editions de Minuit, 1970). See also David Granick, *The European Executive* (New York: Doubleday, 1962), and Ezra N. Suleiman, *Elites in French Society: The Politics of Survival* (Princeton N.J.: Princeton University Press, 1978).

5. Patrick J. Harrigan, *Mobility, Elites, and Education in French Society of the Second Empire* (Waterloo, Ontario: Wilfred Laurier University Press, 1980). See also his "Social Mobility and Schooling in History: Recent Methods and Conclusions," *Historical Reflections*, X, 1 (Spring 1983), pp. 127–141.

6. Peter Lundgreen, "The Organization of Science and Technology in France: A German Perspective," eds. R. Fox and G. Weisz, *The Organization of Science and Technology in France 1808–1914* (Cambridge: Cambridge University Press, 1980), pp. 311–332. Also, *Bildung und Wirtschaftswachstum im Industrialisierungsprozess des 19. Jahrhunderts* (Berlin: Colloquium, 1973). Fritz K. Ringer, *Education and Society in Modern Europe* (Bloomington: Indiana University Press, 1979), pp. 158–162.

7. Harry Paul, "The Issue of Decline in Nineteenth-Century French Science," *French Historical Studies* 7 (1972), pp. 416–450; Joseph Ben-David, *The Scientist's Role in Society, A Comparative Study* (Englewood Cliffs, N.J.: Prentice-Hall, 1971); Robert Gilpin, *France in the Age of the Scientific State* (Princeton, N.J.: Princeton University Press, 1968), pp. 77–123; and Henry Guerlac, "Science and French National Strength," ed. E. M. Earle, *Modern France, Problems in the Third and Fourth Republics* (Princeton, N.J.: Princeton University Press, 1951), pp. 81–105.

8. François Crouzet, "An Annual Index of French Industrial Production in the Nineteenth Century," pp. 245–278, ed. Rondo Cameron, *Essays in French Economic History* (Homewood, Ill.: Irwin, 1970); Maurice Lévy-Leboyer, "Innovation and Business Strategies in Nineteenth- and Twentieth-Century France," pp. 87–135, eds. E. C. Carter, R. Forster, and J. N. Moody, *Enterprise and Entrepreneurs in Nineteenth- and Twentieth-Century France* (Baltimore: Johns Hopkins University Press, 1976); and T. J. Markovitch, *L'Industrie française de 1789 à 1964: Conclusions générales*, vol. 7 in *Histoire quantitative de l'économie française* (Paris: Institut de science économique appliquée, 1966).

9. Charles F. Sabel, *Work and Politics: The Division of Labor in Industry* (Cambridge: Cambridge University Press, 1982). See also Patrick K. O'Brien and Caglar Keyder, *Economic Growth in Britain and France, 1780–1914: Two Paths to the Twentieth Century* (London and Boston: Allen and Unwin, 1978).

10. Michel Foucault, *Surveiller et punir, Naissance de la prison* (Paris: Gallimard, 1975), p. 136–196.

11. *Arts et Métiers* 5 (May 1983), pp. 5–13, 50; 11 (November 1958), p. 121; 6–7 (June–July 1970), p. 21.

12. Monte Calvert, *The Mechanical Engineer in America, 1830–1910, Professional Cultures in Conflict* (Baltimore: Johns Hopkins University Press, 1967), pp. 63–85, 107–139. See also Bruce Sinclair, *Philadelphia's Philosopher Mechanics: A History of the Franklin Institute, 1824–1865* (Baltimore: Johns Hopkins University Press, 1974).

13. J. E. Gerstl and S. P. Hutton, *Engineers: The Anatomy of a Profession, A Study of Mechanical Engineers in Britain* (London: Tavistock, 1966). W. H. G. Armitage, *The Rise of the Technocrats* (London: Routledge and Kegan Paul, 1965). C. Sofer, *Men in Mid-Careers, A Study of British Managers and Technical Specialists* (Cambridge: Cambridge University Press, 1970).

14. F. Ringer, *Education and Society*, p. 19.

15. Terry Shinn, *Savoir scientifique*, pp. 94–98.

16. *Annuaire de la Société des Anciens Elèves des Ecoles d'Arts et Métiers* 2 (1849), pp. 2–4 (hereafter cited as *Annuaire*); John Weiss, *Careers and Comrades: Professional Lives in a French Administration, 1804–1930.* (Unpublished manuscript).

17. H. W. Paul and T. W. Shinn, "The Structure and State of Science in France," *Contemporary French Civilization* 6 (Fall–Winter 1981–1982), pp. 153–193.

18. J. Weiss, "The Lost Baton: The Politics of Intraprofessional Conflict in Nineteenth-Century French Engineering," *Journal of Social History*, XVI (Fall 1982) pp. 3–19.

19. Terry Shinn, *Savoir scientifique*, p. 179.

20. Ibid., pp. 213–214, and Theodore Zeldin, *France 1848–1955: Ambition, Love, and Politics* (Oxford: Oxford University Press, 1973), I, pp. 102–103.

21. Michel Crozier, *The Bureaucratic Phenomenon*, pp. 145–208.

22. Pierre Bourdieu and J. C. Passeron, *Les Héritiers* (Paris: Les Editions de Minuit, 1964).

23. Randall Collins, *The Credential Society: An Historical Sociology of Education and Stratification* (New York: Academic/Harcourt Brace Jovanovich, 1979), pp. 49–72. See also *From Max Weber: Essays in Sociology*, trans. and eds. H. H. Gerth and C. W. Mills (New York: Galaxy, 1958), pp. 57, 240–244.

24. Collins, p. 16.

Chapter 2

1. Terry Shinn, *Savoir scientifique*, p. 66.

2. John H. Weiss, *The Making of Technological Man*, pp. 71–76, 256.

3. Ibid., pp. 75, 90.

4. *Le Monde*, 16 November 1979, pp. 33–34.

5. John Weiss, *Technological Man*, p. 226.

6. Ibid., pp. 241–243.

7. L. M. Leroy, *Vers l'éducation nouvelle* (Paris, 1906), p. 133.

8. Ministère du Commerce, de l'Industrie, des Postes et des Télégraphes, Direction de l'enseignement technique, *L'Enseignement technique en France, Etude publiée à l'occasion de l'exposition de 1900*, 5 vols. (Paris, 1900), vol. I, Paul Buquet, *Etablissements nationaux et écoles supérieures de commerce*, pp. 136–139 (hereafter cited as *L'Enseignement technique en France, 1900*).

9. F. Ringer, *Education and Society*, p. 199.

10. *Le Monde*, 16 November 1979, p. 34.

11. Ibid., pp. 33–35.

12. Frederick Artz, *Technical Education*, p. 101. On descriptive geometry, see also John Weiss, *Technological Man*, pp. 139–143, and René Taton, *L'oeuvre scientifique de Monge* (Paris: Presses Universitaires de France). On the Petite Ecole, see James M. Edmonson, *From Mécanicien to Engineer, Technical Education and the Machine-Building Industry in Nineteenth-Century France*, New York: Garland, 1986).

13. *Technique, Arts, Sciences* 186 (February–March 1965), pp. 7, 17–21 (hereafter cited as *TAS*). See also Robert Fox, "Education for a New Age: The Conservatoire des Arts et Métiers, 1815–1830," in D. S. L. Cardwell, ed., *Artisan to Graduate* (Manchester: Manchester University Press, 1974), pp. 23–38.

14. Arthur Morin, "Notice sur le Conservatoire Impérial d'Arts et Métiers," Ministère de l'Agriculture, du Commerce at des Travaux publics, Commission de l'enseignement professionnel, *Enquête sur l'enseignement professionnel ou recueil de dépositions faites en 1863 et 1864 devant la commission de l'enseignement professionnel*, 2 vols. (Paris, 1865), II, pp. 481–496 (hereafter cited as *L'Enquête sur l'enseignement professionnel, 1865*).

15. *TAS* 186 (February–March 1965), pp. 23–24.

16. Ibid., p. 14, and *TAS* 166 (February 1963), p. 10.

17. Ibid. (September–October 1964), pp. 45–48.

18. Ibid. 166 (February 1963), p. 12.

19. George Weisz, *The Emergence of Modern Universities in France, 1863–1914* (Princeton, N.J.: Princeton University Press, 1983), pp. 41–42.

20. Ibid., p. 225.

21. Terry Shinn, "The French Science Faculty System 1801–1914: Institutional Change and Research Potential in Mathematics and the Physical Sciences," eds. R. McCormmach. L. Pyenson, and R. S. Turner, *Historical Studies in the Physical Sciences* (Baltimore: The Johns Hopkins University Press), pp. 271–332.

22. Harry W. Paul, "Apollo Courts the Vulcans: The Applied Science Institutes in Nineteenth-Century Science Faculties," eds. R. Fox and G. Weisz, *The Organization of Science and Technology in France 1808–1914* (Cambridge: Cambridge University Press, 1980), pp. 155–181 (hereafter cited as *Organization of Science*).

23. Ibid., pp. 163–170.

24. Ibid.

25. Ibid.

26. *Bulletin de la Société des Anciens Elèves des Ecoles d'Arts et Métiers* (hereafter cited as *Bulletin*) 1 (January 1907), pp. 54–55.

27. *Bulletin* 10 (October 1921), p. 625; 1 (January 1927), pp. 15–16.

28. Harry Paul, "The Issue of Decline," *French Historical Studies* 7 (1972), pp. 416–450. See also *The Sorcerer's Apprentice, the French Scientist's Image of German Science 1840–1919* (Gainesville: University of Florida Press, 1972), pp. 161–163; Joseph Ben-David, *The Scientist's Role in Society*, pp. 88–107; Robert Gilpin, *France in the Age of the Scientific State*, pp. 77–123; and Henry Guerlac, "Science and French National Strength," *Modern France*, pp. 81–105. See also Robert R. Locke, *Industrial Development and the Socail Fabric: The End of the Practical Man: Entrepreneurs and Higher Education in Germany, France and Great Britain, 1880–1940* (Greenwich, Conn.: J.A.I. Press, 1984). Locke attacks economic "revisionists" such as Crouzet, Lévy-Leboyer, Markovitch, and Richard Roehl and educational historians such as Fritz Ringer. He concludes that after 1880 Germany forged well ahead of France and England in the field of entrepreneurial performance and higher education. He does not, however, cite Peter Lundgreen or Harry Paul on French and German scientific and technical education.

29. H. Paul, "Issue of Decline," pp. 416–450.

30. Terry Shinn, "The French Science Faculty System," *Historical Studies in the Physical Sciences*, pp. 271–332; George Weisz, *Modern Universities*, p. 206.

31. G. Weisz, *Modern Universities*, pp. 193–194.

32. H. Paul, "Apollo Courts the Vulcans," *Organization of Science*, p. 156.

33. Ibid., pp. 176–177.

34. Ibid., pp. 157, 177.

35. *Arts et Métiers* 4 (May 1983), p. 31.

36. H. Paul and T. Shinn, "The Structure of Science," *Contemporary French Civilization*, p. 164.

37. Fédération des associations et sociétés françaises d'ingénieurs diplômés (FASFID), *Septième enquête socio-économique sur la situation des ingénieurs diplômés* (Octo-

ber 1983), p. 63. Also, "La formation d'ingénieurs en France" (Paris: C.E.F.I., 1979), cited in Luc Boltanski, *Les Cadres: la formation d'un groupe social* (Paris: Les Editions de Minuit, 1982), p. 121.

38. La Documentation française, "Les écoles d'ingénieurs," *Notes et études documentaires* (4045–4047), December 1973. Also, "Ingénieurs et Grandes Ecoles," *Nouvelles de France* 85 (16 January 1981), p. 21.

39. Ibid., p. 25.

40. Ibid., p. 24.

41. Ibid. Also, Georges Ribeill, "Entreprendre hier et aujourd'hui: la contribution des ingénieurs," *Culture Technique, Les Ingénieurs* 12 (March 1984), pp. 77–90.

42. *Arts et Métiers* 1–2 (January–February 1983), pp. 29–31; 6 (July–August, 1983), p. 5. The Offices of the Conférence des Grandes Ecoles and of the Comité national pour le développement des Grandes Ecoles can be found at 60, boulevard Saint-Michel, 75006, Paris.

43. *Arts et Métiers* 2 (March 1983), p. 5; 4 (May 1983), pp. 25–31; and 6 (July–August 1983), p. 5.

44. *Arts et Métiers* (September–October 1983), pp. 36–37.

45. Ibid.

Chapter 3

1. *Archives Nationales* (hereafter cited as *AN*) F 17 (L'Instruction publique), 9820–9827, Ecoles primaires supérieures.

2. *AN* F 17 8708, Principal, Collège de Saverne, 21 October 1863.

3. *AN* F 17 11706–11708, L'Enseignement professionnel, 1850–1880s.

4. *AN* F 17 8701–8731, L'Enseignement secondaire spécial. V. Duruy, *Notes et Souvenirs*, 2 vols. (Paris, 1901): I, p. 253. On Cluny, see *AN* F 17 8732–8752.

5. *AN* F 17 8713, rector of the Academy of Dijon, 12 October 1868.

6. *Notes et Souvenirs*, I, p. 258.

7. C. R. Day, "The Short, Unhappy Life of the Cluny School, 1866–1891," *French Historical Studies* VIII (Spring 1974), pp. 430–431. The *Clunisiens* were recruited from the teacher-training schools and were the sons of village farmers, artisans, shopkeepers, and schoolmasters. They taught mainly in the municipal colleges. See *AN* F 17 8737, 8739, 8742–8747, student files.

8. *Bulletin de l'association amicale des anciens élèves de l'école normale spéciale de Cluny* XIV (April 1884), pp. 54–56, and XVI (April 1886), pp. 37–40. *Revue de l'enseignement spécial* III (15 March 1881), pp. 81–82.

9. *Revue de l'enseignement spécial* I (15 August 1879), pp. 116–117, and II (15 January 1880), p. 9.

10. See the reports of the Conseils de perfectionnement (F 17 8711–8714); the reports of rectors and inspectors (especially F 17 8702 and 8708) and the government inquiry into secondary education of 1864, F 17 6840–6849. C. R. Day, "The Rise and Fall of L'Enseignement Secondaire Spécial, 1865–1902," *Journal of Social History* VI (Winter 1972–1973), p. 186. See also Sandra Horvath-Peterson, *Victor Duruy* (Baton

Rouge: Louisiana State University Press, 1984), and Robert Anderson, *Education in France 1848–1870* (Oxford: Oxford University Press, 1975).

11. A. Corbon, *L'Enseignement professionnel* (Paris, 1859); F. Le Play, *Les ouvriers européens, Etude sur les travaux, la vie domestique et la condition des populations ouvrières de l'Europe* (Paris, 1855).

12. Morin and Tresca, *L'Enseignement industriel*. See also Morin's report, Exposition universelle de Londres de 1862, *Rapports des membres de la section française du jury international sur l'ensemble de l' Exposition* (1862), pp. 151–152, 206–210, 247–248, 262.

13. *Enquête sur l'enseignement professionnel* (1864), I, pp. 399–408.

14. Ministère de l'Agriculture, du Commerce et des Travaux publics, Commission de l'enseignement technique, *Rapport et Notes* (Paris, 1865), p. 85.

15. Morin and Tresca, *L'Enseignement industriel*, pp. 5–6.

16. Georges Ferraud and Félix Martel, "Ecoles primaires supérieures, écoles d'apprentissage, and écoles nationales professionnelles," *Mémoires et documents scolaires*, 2nd series, 9 (Paris: Musée Pédagogique, 1889).

17. *AN* F 17 14348–14356, Ecoles nationales professionnelles, F 17 14350, "Rattachement des écoles professionnelles au Ministère du Commerce, 1880–1906."

18. Rapport présenté à la commission mixte des écoles manuelles d'apprentissage, le 14 octobre 1887, *Mémoires et documents scolaires*, 1st series, 46 (Paris, 1887), pp. 81–84. F. Buisson, *L'Enseignement primaire supérieur et professionnel* (Paris, 1887), p. 24.

19. "Rapport présenté à la commission mixte," *Mémoires et documents scolaires*, p. 109.

20. L. M. Leroy, *Vers l'éducation nouvelle* (Paris, 1906).

21. Eugène Le Chartier, *La France et son Parlement* (Paris, 1912), pp. 387–460. Of 114 deputies who listed technical education and/or who were active in the technical education movement, 78 belonged to conservative parliamentary parties, 23 to the left-center or left, and 13 were independents.

22. The Association française published the *Revue de l'enseignement technique*, 1910–1913, followed by *La formation professionnelle*, 1919–1939, and the *Bulletin de l'enseignement technique*, since 1954.

23. *Bulletin de l'enseignement technique* 1 (8 January 1898), pp. 1–23, and 15 (16 March 1912), pp. 76–79, published by the technical education department of the Ministry of Industry and Commerce, 1898–1914.

24. L. M. Leroy, *L'éducation nationale au XXe siècle* (Paris, 1914), pp. 89–90. *Bulletin de l'enseignement technique* 15 (16 March 1912), pp. 76–79.

25. Ministère du Commerce, de l'Industrie, etc., *L'Enseignement technique en France* (1900), especially Félix Martel, *Ecoles pratiques de commerce et d'industrie*, III, pp. 325–326, 462–465, 548–550.

26. Ibid., III, pp. 451–465 and 529–535.

27. The schools studied were at Agen, Boulogne, Brest, Fourmies, Grenoble, Le Havre, Lille, Le Mans, Mazamet, Narbonne, Reims, Rennes, Saint-Chamond, Saint-Didier-la-Sauve, and Saint- Etienne. Ibid., III, pp. 14–22, 73–75, 95, 128, 153, 180–185, 201, 241, 268–270, 389–393, 409, 474–477, 484, and 528–538. C. R. Day, "Education for the Industrial World: Technical and Modern Instruction in France Under the Third Republic, 1870–1914," *Organization of Science*, pp. 127–153.

28. Ibid., III, p. 534.

29. The schools studied were Agen, Béziers, Boulogne, Brest, Fourmies, Grenoble, Limoges, Le Mans, Mazamet, Reims, Rennes, Saint-Chamond, Saint-Didier, and Saint-Etienne. Ibid., III, pp. 14–22, 42–44, 73–75, 95, 128, 153, 180–185, 226, 241, 268, 389–393, 451–455, 474, 484, and 528.

30. Placide Astier and I. Cuminal, *L'Enseignement technique, industriel et commercial en France et à l'étranger*, 2nd ed. (Paris, 1912), pp. 178–181.

31. *AN* F 17 14348–14356, Ecoles nationales professionnelles. See also Ferraud and Martel, "Ecoles nationales professionnelles," *Mémoires et documents scolaires*, p. 29.

32. F 17 14356 (Nantes), Bayet, director of Primary Education to the Minister of Public Instruction, 31 August 1898.

33. *AN* F 17 14350, Report of M. Leblanc, 1900, and F 17 14354, principal's report, Voiron, 1900.

34. *AN* F 17 14355 (Vierzon), 23 November 1899, and the Société amicale des anciens élèves, 1895.

35. René Leblanc, *La réforme des écoles primaires supérieures* (Paris, 1907), pp. 11–13.

36. *AN* F 17 14350, Statistique graphique des écoles primaires supérieures de garçons, 1901.

37. *Bulletin* 10 (October 1899), pp. 639–649.

38. Astier and Cuminal, *L'Enseignement technique*, pp. 398–405. Ministère du Commerce, de l'Industrie, des Postes et des Télégraphes, Conseil supérieur du Travail, *Enquête récente sur l'enseignement professionnel en France, Rapport de M. Briat au nom de la commission permanente, procès-verbaux des séances* (Paris, 1905), pp. 51–78. See also Ministère du Commerce, etc., Conseil supérieur de l'enseignement technique, *Rapport sur la situation de l'enseignement technique en France en 1904, présenté par Cohendy: procès-verbaux des séances* (Paris, 1905), pp. 1–21.

39. J. P. Guinot, *Formation professionnelle et travailleurs qualifiés depuis 1789* (Paris: Domat, 1946), pp. 175–189.

40. On private schools and the schools under Commerce, see the *Revue de l'enseignement technique* 1 (May 1911), p. 477; 3 (March 1913), p. 283; 6 (March 1913), p. 283. See also Ministère du Commerce, etc., *L'Enseignement technique en France* (1900), vols. 4–5: *Etablissements divers d'enseignement technique*.

41. In 1910, M. Jean Dupuy, minister of Commerce, said that nearly 100,000 young people attended technical and vocational schools and continuing education programs. Leroy, *L'éducation nationale*, p. 86.

42. *Bulletin de l'enseignement technique* 2 (3 June 1899), pp. 263–298.

43. *Bulletin* 45 (October 1909), pp. 791–799, and 38 (September 1902), pp. 705–706. *AN* F 12 6321, Chamber of Commerce of Reims, 14 October 1901.

44. *Bulletin* 11 (November 1906), pp. 846–847.

45. Leroy, *L'éducation nationale*, p. 244.

46. The teachers for the technical schools under Commerce were trained at the Ecole d'arts et métiers de Châlons for industry, at the Ecole des hautes-études commerciales in Paris for business, and at the Ecole pratique of Le Havre for the 14 girls' *écoles*

pratiques, all of which were unified in 1912 into the Ecole normale d'enseignement technique.

Chapter 4

1. *La Formation professionnelle* 23 (1919), p. 472; *Technique, Art, Science* 4–5 (Jan.–Feb. 1955), pp. 4–6.

2. "De l'enseignement technique dans l'enseignement secondaire public de la loi Astier du 25 juillet 1919 à la loi d'orientation de l'enseignement technique du 16 juillet 1971," *TAS* 277–278 (1974), pp. 7–72. See also J. P. Guinot, *Formation professionnelle*, pp. 211–237.

3. John Talbot, *The Politics of Educational Reform in France, 1918–1949* (Princeton, N.J.: Princeton University Press, 1969), pp. 34–64.

4. *L'Enseignement technique* 63 (July-August–Sept. 1969), pp. 55–57.

5. *La Formation professionnelle* 26 (1920), pp. 66–120.

6. A. Prost, *L'Enseignement en France*, pp. 245–269.

7. J. P. Briand, J. M. Chapoulie, and H. Peretz, "Les statistiques scolaires comme représentation et comme activité," *Revue française de sociologie* XX (1979), pp. 669–702.

8. Briand, Chapoulie, and Peretz, "Les conditions institutionnelles de la scolarisation secondaire des garçons entre 1920 et 1940," *Revue d'histoire moderne et contemporaine* XXV (July–Sept. 1979), pp. 391–421.

9. Guinot, *Formation professionnelle*, p. 283.

10. *Bulletin des écoles nationales professionnelles* 70 (July 1928), p. 114.

11. Ibid. 94 (July 1930), p. 109; 167 (August 1936), pp. 186–203; and 7 (January 1947), p. 36.

12. Ibid. 3 (May 1959), pp. 25–28. Société amicale nationale des techniciens supérieurs, brevetés des lycées techniques et des anciens élèves des écoles nationales professionnelles.

13. Guinot, *Formation professionnelle*, pp. 278–284.

14. Institut national de la statistique et des études économiques, *Tableaux de l'économie française* (Paris: I.N.S.E.E., 1978).

15. *L'Enseignement technique* 12 (Oct.–Nov.–Dec. 1956), pp. 10–19.

16. Paul Langevin, "Culture et Humanités," *Pour L'ère nouvelle* (Paris, 1947); and Georges Cogniot, ed., *Le Plan Langevin-Wallon, La nationalisation de l'enseignement* (Paris, 1962).

17. H. Wallon, *De l'acte à la pensée* (Paris, 1942).

18. *L'Enseignement technique* 12 (Oct.–Dec. 1956), pp. 10–19.

19. There were three ENNAs for men at Lyon, Nantes, and Paris and two for women at Paris and Toul, *TAS* (Jan. 1961), pp. 40–42.

20. *L'Enseignement technique* 1 (Jan.–March 1954), pp. 33–38; 12 (Oct.–Dec. 1956), pp. 16–18.

21. Ibid. 6 (April–June 1955), pp. 11–20.

22. *TAS* 19 (April 1948), pp. 12–13.

23. *L'Enseignement technique* 5 (Jan.–March 1955), p. 16.

24. Union des Industries métallurgiques, *Le personnel dans l'industrie des métaux* (Paris: U.I.M.M., 1963). *L'Enseignement technique* 12 (July–Sept. 1958), pp. 20–39.

25. *L'Enseignement technique* 2 (April–June 1954), pp. 58–61; *TAS* 102 (Nov. 1956), pp. 37–41.

26. *L'Enseignement technique* 12 (Oct.–Dec. 1956), pp. 33–34.

27. J. N. Moody, *French Education*, p. 196.

28. *L'Enseignement technique* 61 (Jan.–March 1969), p. 58.

29. *Bulletin des écoles professionnelles* 3 (May 1958), pp. 25–28; 15 (May 1961), p. 33.

30. *L'Enseignement technique* 57 (Jan.-March 1968), pp. 21–28; 69 (Jan.–March 1971), p. 24.

31. Ibid. 63 (July–Sept. 1969), pp. 53–56.

32. Ibid. 61 (Jan.–March 1969), p. 58.

33. Ibid. 64 (Oct.–Dec. 1969), pp. 34–38.

34. J. N. Moody, *French Education*, p. 205.

35. *L'Enseignement technique* 69 (Jan.–March 1971), p. 21.

36. Ibid. 71 (July–Sept. 1971), pp. 8–14; 72 (Oct.–Dec. 1971), p. 21.

37. Ibid. 70 (April–June 1971), pp. 11–12.

38. Ibid. 86 (April–June 1975), p. 99.

39. Ibid. 110 (April–June 1981), pp. 66–99. See also *Le Monde de l'Education* 2 (January 1975), p. 3, and René Haby, *Propositions pour une modernisation du système éducatif* (Paris: La Documentation française, 1975).

40. *L'Enseignement technique* 110 (April–June 1981), p. 99.

41. Ibid. p. 66.

42. *Nouvelles de France* 85 (16 Jan. 1981), pp. 24–25.

43. "Réussir le Bac," *Le Monde de l'Education* 92 (March 1983), pp. 17–23; Pierre Bourdieu, *La Distinction* (Paris: Editions de Minuit, 1979), pp. 172–173.

44. Alain, *Propos sur l'éducation* (Paris: Presses Universitaires de France, 1963), pp. 57–70; *TAS* 279–280 (1974), pp. 3–5.

45. *TAS* 134 (December 1959), p. 9, and (January 1961), pp. 13–25. See also Antoine Léon, *Formation générale et apprentissage du métier* (Paris: Presses Universitaires de France, 1965), pp. 30–35.

46. *TAS* 7 (April 1950), pp. 1–3.

47. H. Luc, "Les problèmes actuels de l'enseignement technique," *L'Enseignement technique* 1 (1938), p. 35.

48. H. Wallon, *De l'acte à la pensée* (Paris, 1942).

49. G. Friedmann, *Industrial Society* (Glencoe, Ill.: Free Press, 1955), p. 186.

50. A. Sauvy, *La montée des jeunes* (Paris: Calmann-Lévy, 1959). *TAS* 145 (January 1961), pp. 13–25, and 174 (December 1963), pp. 8–15.

51. S. Mallet, *La nouvelle classe ouvrière* (Paris: Editions du Seuil, 1963, 2nd ed. 1969).

L. Boltanski, *Les Cadres* (Paris: Editions de Minuit, 1982), pp. 269–275. See also the works of P. Gorz, Pierre Naville, and Alain Touraine.

52. Fédération nationale des anciens élèves des enseignements techniques et professionnels, "La situation actuelle et l'avenir de l'enseignement technique," *L'Education professionnelle* 41 (1964), pp. 13–20.

53. *L'Enseignement technique* 109 (Jan.–March 1981), pp. 17–48.

Chapter 5

1. André Prévot, *L'Enseignement technique chez les Frères des Ecoles Chrétiennes au XVIIIe et XIXe siècles* (Paris: Ligel, 1932), pp. 5–106.

2. F. Artz, *Technical Education*, pp. 77–81.

3. Camille Ferdinand-Dreyfus, *Un philanthrope d'autrefois, La Rochefoucauld-Liancourt, 1747–1827* (Paris: Plon, 1903), pp. 371–413. See also Jean-Dominique de La Rochefoucauld, Claudine Wolikow, and G. Ikni, *Le Duc de La Rochefoucauld-Liancourt de Louis XV à Charles X, Un grand seigneur patriote et le mouvement populaire* (Paris: Perrin, 1981).

4. *AN* F 12 1085, La Rochefoucauld to the Minister of the Interior, 15 January 1807.

5. Ferdinand-Dreyfus, p. 372.

6. *Annuaire* I (1848), pp. 15–32.

7. F. Le Brun, "Notice sur les écoles impériales d'Arts et Métiers," *Enquête sur l'enseignement professionnel* 2 (1865), pp. 581–605.

8. *AN* F 12 1084, Etat des élèves de l'école d'arts et métiers de Compiègne, February 1804.

9. F. Euvrard, *Historique de l'Ecole nationale d'arts et métiers de Châlons-sur-Marne 1780–1895* (Châlons-sur-Marne: Union Républicaine, 1895), pp. 6–7.

10. *AN* F 12 1105, Note to Montalivet, minister of the Interior, n.d., around 1810.

11. *AN* F 12 1084, 2 Sept. 1808; F 12 1085, 4 Dec. 1806; and 15 Jan. 1807.

12. *AN* F 12 1085, La Rochefoucauld to the minister of the Interior, 24 May 1808.

13. Ibid., 15 January 1807.

14. *AN* F 12 1084, 8 November 1808.

15. Ibid.

16. Ibid., 15 January 1807.

17. Ibid., September 1806.

18. Ibid.

19. Ibid., 15 January 1807.

20. Ibid., 11 July 1806.

21. Ibid.

22. *Annuaire* 1 (1848), pp. 6, 23.

23. *AN* F 12 1098, 4 November 1814.

24. Paul Popin, *Les Gadzarts* (Saint-Dizier: Ets. Brulliard, 1947), p. 50.

25. *AN* F 12 1084, correspondence April 1817.

26. *AN* F 12 1105, Decree of 27 February 1817; La Rochefoucauld to the minister, 9 May 1818. André Guettier, *Historie des Écoles nationales d'arts et métiers* (Paris: Dejay, 2nd ed. 1880), p. 40.

27. La Rochefoucauld was a director of the Société d'encouragement pour l'industrie nationale; the Société pour l'éducation élémentaire, founded in 1816 to encourage the new Lancaster (mutual) teaching method in France; the Conservatoire des arts et métiers of Paris; and various societies for the reform of hospitals, prisons, and insane asylums. See Camille Ferdinand-Dreyfus, *Un philanthrope d'autrefois*, pp. 455–469.

28. Euvrard, *Châlons*, pp. 61–62.

29. *AN* F 12 1167, Jeandeau (acting director), 26 August 1830.

30. *AN* F 12 1139A, 20 August 1819.

31. *AN* F 12 1156, 10 September 1825; 1167, report of 1831.

32. *Annuaire* 1 (1848), pp. 6, 23–24; 4 (1851), p. 60.

33. A. Guettier, *Ecoles d'arts et métiers*, pp. 54, 350–352.

34. Ibid.

35. *AN* F 12 4879, report of A. Vincent, 25 June 1842.

36. Ibid.

37. *AN* F 12 1168, 21 August 1833; F 12 1170 (Drôme), 21 August 1833; F 12 1171 (Loiret), 18 September 1834; and 1172 (Nord), 30 August 1834.

38. *AN* F 12 4879, 25 June 1842.

39. Raymond Grew, Patrick Harrigan, James Whitney, "La scolarisation en France, 1829–1906," *Annales: Economies, Sociétés, Civilisations* 1 (January–February 1984), pp. 116–157.

40. *AN* F 12 1171, 28 August 1840.

41. *AN* F 12 1171, report of 1840; F 12 1172, 13 December 1837 and 31 August 1839.

42. J. Vial, *L'Industrialisation de la sidérurgie, 1814–1864* (Paris: Mouton, 1967), pp. 47, 201–202, 236–237.

43. *AN* F 17 14335, Aix-en-Provence, 1837–1856, Conseil général des Bouches-du-Rhône, session of 2 September 1841.

44. Ibid. and F 12 1169, Prefect Bouches-du-Rhône to minister of Commerce, 30 September 1840.

45. *Ibid.*, Réponse à la circulaire du 14 juillet 1838 par laquelle les départements du midi ont été consultés sur le lieu le plus convenable pour l'établissement d'une 3e Ecole.

46. Ibid., ministerial note of 23 October 1841.

47. Vial, *L'Industrialisation de la sidérurgie*, p. 196.

48. AN F 17 14335, minister of Commerce to Thiers, 16 June 1842.

49. *Annuaire* 19 (1866), pp. 107–108.

50. Ibid. 4 (1851), pp. 16–55, Extraits du Moniteur universel, 22 and 26 July 1850.

51. Ibid.

52. A. Guettier, *Ecoles d'arts et métiers*, pp. 73–81, 360–366.

53. *Annuaire* 3 (1850), pp. ix–x. Of 1343 graduates traced in 1850, 907 (67 percent) worked in industry and transport, 185 (14 percent) in the military, 164 (12 percent) in state services and teaching, 34 as architects and artists, and 53 as miscellaneous.

54. Extrait du Moniteur universel, séance du 26 avril 1850, in Guettier, *Ecoles d'arts et métiers, pp. 356–374.*

55. Extrait du Moniteur universel, séance du 22 juillet 1850, *Annuaire* 4, pp. 16–25.

56. *Annuaire* 12 (1859), pp. 16–17.

57. A. Guettier, "Etude sur l'instruction industrielle," *Bulletin* 17 (January–March 1864), pp. 21–48; "De la propagation des connaissances industrielles," *Bulletin* 18 (April–June 1864), pp. 15–52.

58. *Annuaire* 10 (1857), pp. 207–213.

59. *Bulletin* 214 (September 1882), pp. 607–612; 2 (February 1892), pp. 143–154; (December 1933), pp. 301–302. The school years indicated reflect the French method of measuring by the first year of studies (*promotion*); hence Fontaine started at Châlons in 1848 and graduated in 1851.

60. *Annuaire* 19 (1866), pp. 110–113.

61. Ibid.

62. *Bulletin* 137 (May 1876), pp. 217–246; 168 (November 1878), pp. 582–589.

63. *Annuaire* 2 (1849), pp. 4, 39–48.

64. Weiss, *Careers and Comrades*, pp. 174–194; *Bulletin* 70 (October 1869), pp. 359–361.

65. Ibid.

Chapter 6

1. Jean Primault et al., *Le Livre d'Or du Bicentenaire des Ecoles d'Arts et Métiers* (Paris: Arts et Métiers, 1980), pp. 741–748.

2. Ibid., p. 747. See also the *Bulletin* 189 (August 1880), pp. 420–424.

3. *Bulletin* 183 (February 1880), p. 96; 189 (August 1880), pp. 434–435.

4. Ibid. 225 (August 1883), pp. 378–384.

5. Guettier, *Histoire des Ecoles d'arts et métiers*, p. 210.

6. *Bulletin* 188 (July 1880), pp. 376–384; 224 (July 1883), pp. 316–324.

7. *Le Siècle*, 10 December 1879; *Bulletin* 179 (Dec. 1879), pp. 716–719.

8. *Bulletin* 179 (Dec. 1879), pp. 716–719.

9. Ibid. 172 (March 1879), p. 173.

10. Ibid. 245 (April 1885), p. 270.

11. D. Poulot, *Le Sublime, ou le travailleur comme il est en 1870 et ce qu'il peut être* (Paris, 1872), reissued in 1982 by Editions Maspéro, Alain Cottereau, editor.

12. *Arts et Métiers* 1 (Jan.–Feb. 1983), p. 50.

13. D. Poulot, *Manifeste d'un Bourgeois démocrate* (Paris, 1871), p. 15.

14. Ibid., p. 17.

15. *Bulletin* 187 (June 1880), pp. 311–319; 217 (June 1882), p. 443.

16. Ibid. 216 (November 1882), pp. 775–785.

17. Ibid. 202 (September 1881), pp. 490–494.

18. *Manifeste*, p. 17.

19. *Bulletin* 207 (January 1882), pp. 65–71.

20. Ibid. 234 (May 1884), pp. 306–309.

21. Ibid. 231 (February 1884), pp. 135–140.

22. Ibid. 245 (April 1885), pp. 249–252.

23. Ibid. 246 (May 1885), pp. 330–331.

24. Ibid. 247 (June 1885), pp. 375–381.

25. Ibid. 9 (September 1885), pp. 517–518; 12 (December 1885), pp. 718–719.

26. Ibid. 168 (November 1878), p. 580.

27. *AN* F 17 14348–14356, Ecoles Nationales Professionnelles.

28. *Bulletin* 3 (March 1898), pp. 295–312.

29. *AN* F 17 14319, Note sur les principales améliorations apportées au régime des Ecoles d'Arts et Métiers, 26 February 1901. See also Millerand's speeches to the association, *Bulletin* 2 (February 1900), pp. 164–187, and 2 (February 1901), pp. 138–144.

30. *Annuaire* 52 (1900), pp. 35–52. In France there were 83 branches in 66 departments, 8 in the colonies, and 15 in foreign countries.

31. *Bulletin* 2 (February 1903), pp. 150–175. The ratio of those placed annually by the association's placement service to the number of job seekers registered with it was 99/320 in 1883, 365/931 in 1900, and 510/1200 in 1911. Those unemployed, as opposed to those seeking a change in position, rose from a monthly average of 30 to 80 in the 1890s to 200 to 300 between 1900 and 1905. See the annual February reports of the placement committee in the *Bulletin* and the study of the years 1883–1925 in 5 (May 1925), p. 329.

32. The number of candidates rose from 1,000 in 1881 to 1,264 in 1892 and to about 1,500 in the years before the war. Three fourths of the students had full or partial aid from the government. *Bulletin* 205 (December 1881), p. 764; 2 (February 1892), p. 105; and 6 (June 1914), p. 687.

33. André Prévot, *L'enseignement technique chez les frères*, p. 135.

34. *AN* F 17 14337 (Lille), petition dated 12 May 1879, speech of M. Tirard, 1 June 1879.

35. Ibid., and *Bulletin* 12 (December 1902), pp. 861–863.

36. *AN* F 17 14341 (Cluny). See also F 17 14345 (Paris), Projet de création, 1901–1905.

37. *Bulletin* 8 (August 1901), pp. 504–507, 570–573; 10 (October 1901), p. 668; 6 (June 1902), pp. 419–426; and 10 (October 1902), pp. 822–825.

38. *Bulletin* 6 (June 1902), pp. 424, 499–502.

39. Ibid. 1 (January 1908), pp. 45–46; 8–9 (August–September 1909), pp. 692, 718. AN F 17 14340, director of Lille to the minister, 12 August 1919. In 1914 the five schools produced 422 graduates, of which 262 (62 percent) won the engineering diploma.

40. Ibid.

41. H. Paul, in R. Fox and G. Weisz, eds., *Organization of Science*, pp. 158–171.

42. *Bulletin* 8–9 (August–September 1912), pp. 814, 822.

43. *AN* F 17 14345 (Paris), reports of 1901 and 1905 on the Ecoles d'arts et métiers. For the years 1898 to 1902, of 1,611 graduates, 1,206 (75 percent) went into industry and transport, 227 (14 percent) into higher education, and 178 (11 percent) unknown. On scholarships to the Ecole centrale, see the *Bulletin* 3 (March 1887), pp. 156–158.

44. Morin, *L'enseignement industriel*, p. 30.

45. *Bulletin de l'enseignement technique* VIII (14 April 1906), p. 153, and F 17 14345 (Paris).

46. *Bulletin* 1 (January 1927), pp. 15–16.

47. Ibid. 10 (October 1909), p. 966.

48. Ibid. 10 (October 1912), p. 1028.

49. Ibid. 9 (September 1921), pp. 630–632.

50. Ibid. 1 (January 1931), p. 69.

51. P. Popin, *Les Gadzarts*, p. 81. See also, A. Grelon, "L'Education des cadres: La question des aspirations professionnelles chez les futurs cadres d'entreprise," thèse de troisième cycle, Université de Paris VII, 1983.

52. F 17 14319, Groupe des Ingénieurs des Arts et Métiers de Boulogne-sur-Seine, November 1918. See also the article by Jean Fieux in *Technique, Art, Science* (February 1949), pp. 1–6.

53. Aimée Moutet, "Ingénieurs et rationalisation dans l'industrie française de la Grande Guerre au Front Populaire," *Culture Technique* 12 (March 1984), pp. 137–153. See also Edwin T. Layton, Jr., *Revolt of the Engineers: Social Responsibility and the American Engineering Profession* (Cleveland: Case Western Reserve University Press, 1971), pp. 30–31, 134–149.

54. *Bulletin* 2 (February 1885), p. 129; 2 (February 1902), pp. 158–161; 1 (January 1930), pp. 56–61. Bruno Jacomi, "La Société des Ingénieurs civils, 1848–1983," 12 *Culture Technique*, pp. 209–219. See also, La Société des ingénieurs civils de France, *125 ans de progrès technique vus à travers la société des ingénieurs civils de France* (Paris: ICF, 1973).

55. Félix Faure inaugurated the new headquarters at the rue Chauchat in 1895. *Bulletin* 2 (February 1895), p. 92. See also the account of President Loubet's visit, 2 (February 1902), pp. 171–174.

56. Ibid. 2 (February 1898), p. 198.

57. Ibid. 7 (July 1892), pp. 401–407.

58. Ibid. 206 (January 1882), p. 33.

Chapter 7

1. *Bulletin* 9–10 (September–October 1921), pp. 610–615.

2. Ibid. 11 (November 1920), pp. 591–593; 12 (December 1921), pp. 806–808.

3. *La Formation professionnelle* 26 (1920), pp. 66–127.

4. *Bulletin* 2 (February 1921), pp. 131–137, decree of 7 February 1921.

5. Ibid. 8–9 (August–September 1909), pp. 749–763; 2 (February 1923), pp. 167–174.

6. *Arts et Métiers, Revue technique mensuelle* 120 (September 1930), pp. 361–363.

7. *Ingénieurs Arts et Métiers* 2 (February 1938), p. 102.

8. *Bulletin* 5 (May 1929), pp. 430–435; 8 (August 1929), p. 863.

9. P. Popin, *Les Gadz'arts*, pp. 150–152.

10. *Bulletin* 12 (December 1920), pp. 826–828; *Ingénieurs Arts et Métiers* 10 (October 1938), p. 128.

11. Edmond Labbé, *Les outils de perçage, leçon type de technologie, faite à des élèves de cours professionnels* (Paris, 1926).

12. *Bulletin* 12 (December 1921), p. 805.

13. Popin, pp. 178–179.

14. Labbé, *Les outils de perçage*, p. 2, and *Bulletin* 8 (August 1927), pp. 646–651.

15. *Bulletin* 12 (December 1924), pp. 927–929, 939; 2 (February 1927), pp. 127–135; and Popin, *Gadz'arts*, pp. 161–163.

16. *Bulletin* 8 (August 1921), pp. 522–525; 2 (February 1932), p. 114.

17. Ibid. 3 (March 1926), p. 156.

18. Ibid. 10 (October 1930), p. 928.

19. Ibid. 12 (December 1935), p. 851.

20. Luc Boltanski, *Les Cadres*, pp. 66, 94–125. See also André Thépot, ed., *L'ingénieur dans la société française* (Paris: Editions ouvrières, 1985), and André Grelon, "L'éducation des cadres," 1983.

21. Alfred Chandler, *The Visible Hand: The Managerial Revolution in American Business* (Cambridge, Mass.: Harvard University Press, 1977), and A. Chandler and Herman Daems, eds., *Management Hierarchies: Comparative Perspectives on the Rise of Modern Industrial Enterprise* (Cambridge, Mass.: Harvard University Press, 1980).

22. *Bulletin* 9 (September 1934), pp. 547–552.

23. *Bulletin* 9 (September 1931), p. 843; 6 (June 1933), p. 406; 2 (February 1936), p. 108. *Ingénieurs Arts et Métiers* 10 (October 1938), p. 127; 3 (March 1942), p. 11.

24. *Bulletin* 12 (December 1933), p. 799.

25. Ibid. 6 (June 1933), p. 408. *Ingénieurs Arts et Métiers* 5 (May 1938), p. 4.

26. *Ingénieurs Arts et Métiers* 2 (February 1938), pp. 102, 143–145; 5 (May 1938), p. 5.

27. Ibid. 7–8 (July–August 1938), pp. 123–129.

28. *Bulletin* 12 (December 1935), p. 855.

29. *Ingénieurs Arts et Métiers* 7–8 (July–August 1938), pp. 125–130; 2 (February 1940), p. 18.

30. Ibid. 11 (November 1938), p. 155; 4 (April 1940), p. 57.

31. Ibid. 7–8 (July–August 1938), pp. 180–183.

32. *Bulletin* 12 (December 1933), p. 800.

33. *Ingénieurs Arts et Métiers, Techniques et Variétés* 10 (October 1938), pp. 179–184.

34. L. Boltanski, *Les Cadres*, pp. 83–84.

35. *Ingénieurs Arts et Métiers* 237 (December 1941), pp. 1–2; 234 (March 1941), pp. 1–2; 238 (March 1942), pp. 7–8.

36. Ibid. 237 (December 1941), p. 1.

37. Ibid. 254 (December–January 1945–1946), p. 11. *Technique, Art, Science* (January 1956), pp. 7–17.

38. *Bulletin* 2 (February 1937), p. 91; 1 (January 1931), p. 31.

39. *Ingénieurs Arts et Métiers* 245 (October 1943), p. 4. *TAS* (February 1949), pp. 1–6.

40. Ibid. 254 (December–January 1945–1946), p. 6.

41. *Arts et Métiers* 6 (June 1956), p. 45; 6 (June 1957), pp. 28–34.

42. *Revue de l'enseignement technique* 12 (October 1956), p. 21.

43. Union des industries métallurgiques et minières, de la construction mécanique, électrique et métallique et des industries qui s'y rattachent, *Ingénieurs et cadres supérieurs, situation actuelle et prévision des besoins dans les industries des métaux* (Paris: U.I.M.M., 1956), pp. 12–15, 18, 45–51.

44. *Arts et Métiers* 11 (November 1958), pp. 117–119.

45. Ibid. p. 121.

46. Ibid. 4 (April 1963), p. 97.

47. Ibid. 6–7 (June–July 1970), p. 21; 5 (May 1983), pp. 5–13, 23; 11 (November 1958), p. 121. In 1970, the Arts et Métiers graduated 600 per year of a total of 8,000 graduating engineers in France, or 7.5 percent; in 1982 they graduated 720 of 15,000, or 5 percent.

48. *Arts et Métiers* 12 (December 1962), p. 61; 1 (January 1964), p. 33.

49. Ibid. 1 (January 1956), p. 13.

50. Ibid. 11 (November 1958), p. 121.

51. Ibid. 4 (April 1963), p. 67.

52. Ibid. 7–8 (July–August 1967), pp. 28–31; 2 (February 1968), pp. 11–13.

53. Ibid. 3 (March 1966), p. 67.

54. Ibid. 12 (December 1970), pp. 11–13.

55. Ibid. 4 (April 1963), p. 67.

56. Ibid. 7 (July–August 1967), pp. 29, 54; 6 (July–August 1983), p. 5.

57. Ibid. 6 (June–July 1970), p. 23.

58. Ibid. pp. 21–26.

59. H. Paul and T. Shinn, "The Structure of Science," pp. 153–193.

60. Ibid.

61. *Arts et Métiers* 10 (December 1980), p. 24; 4 (May 1983), pp. 37–40.

62. Ibid. 4 (May 1983), pp. 52–55.

63. Ibid. 4 (May 1982), pp. 19–27.

64. Ibid. 1 (January–February 1982), pp. 29–30.

65. Ibid. 4 (May 1983), p. 52, and Société des Ingénieurs Arts et Métiers, "Sondage sur l'origine sociale des parents et grands-parents des élèves actuellement en 4ème année de l'ENSAM," 1975.

66. *Arts et Métiers* 7 (July–August 1975), pp. 33–35.

67. *Livre d'Or*, p. 768.

68. Ibid., pp. 769–773.

69. E. Plenel, "Les mystères des Arts et Métiers," *Le Monde* 3–4 (March 1982).

70. *Arts et Métiers* 4 (May 1983), p. 13.

71. Ibid. 2 (March 1983) p. 25; 1 (January–February 1983), p. 30.

72. Ibid. 7 (Sept.–Oct. 1983), pp. 36–37.

Chapter 8

1. Emile Durkheim, *L'évolution pédagogique en France* (Paris: Alcan, 1938), pp. 140–143.

2. Michel Foucault, *Surveiller et punir*, pp. 136–196.

3. *AN* F 12 1084, La Rochefoucauld to minister of the Interior, 19 March 1807; F 12 1085, 15 January 1807.

4. *AN* F 12 1085, La Rochefoucauld to minister of the Interior, 6 March 1807.

5. Ibid.

6. *AN* F 12 1085, Labâte to minister of the Interior, 8 July 1809, 14 April 1809.

7. Ibid., La Rochefoucauld to minister of the Interior, 6 March 1807.

8. Ibid., Labâte to minister of the Interior, 23 April 1806; prefect Department of the Marne, 1 March 1809.

9. *AN* F 12 1084, La Rochefoucauld to minister, 19 March 1807; 1085, 6 March 1807, and 24 May 1808.

10. Ibid.

11. Ibid.

12. *AN* F 12 1804, 13 March 1807.

13. Ibid., 3 November 1809, 24 May 1809.

14. F 12 1085, undated, probably September 1806.

15. F. Euvrard, *Châlons*, p. 63.

16. Paul Gélineau, *Gadz'arts* (Paris: the author, 1910), p. 183.

17. *AN* F 12 1156, Boisset to minister of the Interior, 10 Sept. 1825.

18. *AN* F 7 6978, report of Captain Barthélemy, commander of the Marne gendarmerie, 3 April 1826.

19. A. G. Claveau, *De la Police générale et de ses abus* (Paris, 1831), p. 464.

20. *AN* F 12 1152, Boisset to minister of the Interior, 19 February 1827.

21. Ferdinand-Dreyfus, *La Rochefoucauld*, pp. 368–370.

22. *AN* F 12 1197, Etats mensuels, 1825–1827; 1167, Jeandeau (acting director), 26 August 1830.

23. F. Euvrard, *Châlons*, pp. 82–84.

24. *AN* F 12 4879.

25. John Weiss, *Technological Man*, p. 183.

26. Jules Guillou, A. Chalain, C. Serre, and A. Voyer, *Un siècle sous les cloîtres du Ronceray, Ecole nationale des arts et métiers d'Angers* (Angers: the authors, 1956), pp. 30–38.

27. *AN* F 12 4875, Inspector Le Brun, reports of 1856, 1861–1862, and 1864–1866.

28. F. Euvrard, *Châlons*, pp. 149–167.

29. *AN* F 17 14340 (Lille), M. Vellutini, November 1879.

30. Paul Popin, *Les Gadz'arts*, p. 156.

31. P. Nizan, *Antoine Bloyé* (New York: Monthly Review Press, 1973), p. 59, originally published in Paris: Grasset, 1933.

32. J. Gillou et al., *Angers*, p. 106.

33. P. Gélineau, *Gadz'arts*, p. 185.

34. Ibid., pp. 27–29, 44.

35. Ibid., pp. 157–178.

36. Gélineau, pp. 27–29.

37. J. Guillou et al., pp. 30–31.

38. Ibid., p. 81.

39. Euvrard, p. 200.

40. Gélineau, p. 94.

41. *Bulletin* 3 (March 1898), pp. 296–300.

42. H. Labarde, *Le Ronceray, Ecole de Gadz'arts, 1527–1927* (Pau: the author, 1941), p. 114.

43. *Bulletin* 4 (April 1900), p. 260. See also *Le Gadzarts, bimestriel d'information de l'union des élèves de l'ENSAM* 7 (January–February 1983), pp. 40–42.

44. *AN* F 17 14335 (Aix), 12 May 1896.

45. Gélineau, pp. 221–222.

46. *AN* F 17 14319, "Note sur les mutineries," November 1894 and June 1898.

47. Emile Hinzelin, "Les Ecoles d'Arts et Métiers," *Manuel général de l'Instruction publique* 38 (21 June 1902), pp. 385–386.

48. Gélineau, p. 187.

49. *Bulletin* 3 (March 1898), pp. 308–309.

50. *AN* F 17 14319, especially "Relevés et notes sur les mutineries et les exclusions pendant les dernières années," 7 April 1899.

51. Guy Dallay *Qu'est-ce qu'un Gadz'arts?* (Talence: the author, 1943), p. 25.

52. Denys Cuche, "Traditions populaires ou traditions élitistes? Rites d'initiation et rites de distinction dans les Ecoles d'Arts et Métiers," paper delivered at the conference Les Cultures Populaires, Université de Nantes, 9–10 June 1983.

53. Gélineau, p. 224.

54. Ibid., p. 185.

55. Ibid., pp. 245–247.

56. J. Guillou et al., p. 112.

57. Gélineau, p. 250.

58. Ibid., pp. 286–290; Euvrard, pp. 225–226.

59. Gélineau, pp. 186–187, 217–218, 290–297.

60. Ibid., pp. 50–61, 228–239, 281–283.

61. Ibid., pp. 44–45, 225–226.

62. Jesse R. Pitts, "Continuity and Change in Bourgeois France," in Stanley Hoffmann et al., *In Search of France* (New York: Harper and Row, 1963), pp. 235–304.

63. *Bulletin* 2 (February 1899), pp. 91–92; 247 (June 1885), p. 348.

64. Gélineau, p. 93.

65. Ibid., p. 184.

66. Ibid., pp. 221, 333.

67. Cuche, "Traditions populaires," p. 5.

68. *Bulletin* 1 (January 1909), p. 72.

69. Gélineau, p. 270.

70. Ibid., p. 183.

71. Ibid., pp. 195–196.

72. Ibid., pp. 111–114.

Chapter 9

1. *AN* F 12 4879, 25 June 1842.

2. *AN* F 12 4875, annual reports of 1855 and 1856.

3. Ibid., report of 1864–1865.

4. *AN* F 17 14340 (Lille), report of 30 March 1904.

5. *Bulletin* 3 (March 1898), pp. 300–312.

6. AN F 17 14319, Ministère du Commerce, de l'Industrie, des Postes et des Télégraphes, Commission chargée d'étudier les améliorations à apporter au régime moral et disciplinaire des écoles nationales d'arts et métiers, Paris, 1899, p. 1.

7. Ibid., p. 11.

8. E. Hinzelin, *Manuel général* (1902), p. 385. See also F 17 14319.

9. *Bulletin* 10 (October 1908), p. 903.

10. Ibid. 2 (February 1912), p. 212.

11. Ibid. 2 (February 1898), pp. 199–200. Emile Gautier, editor of *La Science française*, called them "le ferment qui fait lever la pâte nationale."

12. P. Nizan, *Antoine Bloyé*, pp. 48–49.

13. *Bulletin* 2 (February 1930), p. 160.

14. *Antoine Bloyé*, p. 111.

15. Ibid., p. 234.

16. Ibid., p. 113.

17. W. D. Redfern, *Paul Nizan, Committed Literature in a Conspiratorial World* (Princeton, N.J.: Princeton University Press, 1972), pp. 9–11.

18. *Bulletin* 164 (July 1878), p. 387; 7 (July 1899), pp. 418–421.

19. *Bulletin* 20 (Oct.–Nov.–Dec. 1864), p. 10.

20. Gélineau, p. 270.

21. Ibid., p. 92.

22. Ibid., p. 271.

23. In 1923 only 120 of 11,000 association members had five or more children, *Bulletin* 8 (August 1923), p. 683. It is not unusual, however, to find families that have produced three or four generations of gadzarts: 1 (Jan.–Feb. 1982), p. 58.

24. David Granick, *The European Executive* (New York: Doubleday, 1962), p. 326.

25. Gélineau, p. 272.

26. *Bulletin* 9 (Sept. 1900), pp. 660–664. Georges Duveau, *La pensée ouvrière sur l'éducation pendant la Seconde République et le Second Empire* (Paris: Domat, 1948), pp. 270–271.

27. *AN* F 17 14319, Conférence du groupe des ingénieurs arts et métiers de Boulogne-sur-Seine, November 1918.

28. M. Crozier, *Bureaucratic Phenomenon*, pp. 180–184; L. Wylie, *Village in the Vaucluse* (Cambridge, Mass.: Harvard University Press, 1974), pp. 55–133; J. Pitts, "Bourgeois France," pp. 235–304.

29. Crozier, pp. 180–184.

30. *Bulletin* 242 (January 1885), pp. 41–42.

31. Ibid. 231 (February 1884), pp. 68–92. In 1934 the association sent a trusted vice-president as its representative on a ministerial commission investigating troubles at Angers. He sided with the school officials in blaming the students and calling for appropriate punishment. This unheard of act sparked a revolt among the younger *promotions*, and in the following election to the executive, the vice-president finished eleventh of twelve candidates, defeating only M. de Mégève, who was 85 percent "mutilé de guerre." *Bulletin* 1 (January 1935), pp. 41–44.

32. *Bulletin* 2 (February 1900), pp. 164–167.

33. AN F 17 14319, Note sur les principales améliorations apportées au régime des Ecoles d'arts et métiers, 26 February 1901.

34. Ibid. 9 (October 1909), p. 810, Lettre adressée par le Président de la Société aux élèves des trois divisions de chacune des cinq Ecoles nationales d'arts et métiers, 10 September 1909.

35. A. Metton, *Les Gadz'arts, Les Ingénieurs des Arts et Métiers* (Paris, 1925), p. 152. Metton was the secretary-general of the society. See also Jean Primault et al., *Livre d'Or*, 1980, pp. 417–455.

36. *Annuaire et Liste générale des anciens élèves* 52 (1900), p. 12.

37. *Bulletin* 243 (February 1885), pp. 127–128.

38. Ibid. 8–9 (August–September 1910), pp. 736–742.

39. Guillou et al, *Ronceray, p. 109.*

40. Guillou et al, pp. 111–112.

41. Gélineau, *Gadz'arts*, pp. 111–114; Popin, pp. 110–114.

42. *Arts et Métiers* 11 (November 1964), pp. 21–24.

43. Le Gadz'Arts, *Bimestriel d'information de l'union des élèves de l'ENSAM* 7 (January–February 1983), pp. 9, 15, 38.

44. E. Plenel, "Les mystères des Arts et Métiers," *Le Monde* 3 and 4 (March 1982).

45. *Le Gadz'Arts*, pp. 9, 15, 36–38.

46. *Arts et Métiers* 4 (May 1983), pp. 52–53; *Le Gadz'Arts*, p. 15.

47. *Arts et Métiers* 4 (April 1937), pp. 263–264; 1 (January–February 1983), p. 50, and interviews in Paris in 1979 and 1983.

48. Ibid. 11 (November 1978), p. 53.

Chapter 10

1. *Liste générale alphabétique et par promotions des anciens élèves des Ecoles nationales d'arts et métiers depuis leurs fondations* (Paris: A & M, 1900); *Liste générale décennale et par promotions des anciens élèves des Ecoles nationales d'arts et métiers, 1922–1923* (Paris: Chaix, 1923). See also *AN* F 12, 1084–1086, 1090– 1110, 1135–1187. C. R. Day, "The Making of Mechanical Engineers," *French Historical Studies*, pp. 451–458.

2. *Liste générale 1922–1923*, pp. v–xx; *Bulletin* 2 (Feb. 1937), p. 91.

3. Michel Bouillé, *énseignement technique et idéologies au XIXe siècle*, thesis, Ecole pratique des hautes etudes, Paris, 1972.

4. Société des Ingénieurs Arts et Métiers, "Sondage sur l'origine sociale des parents et grand-parents des élèves actuellement en 4ème année de l'ENSAM," 1975.

5. F. Ringer, *Education and Society in Modern Europe*, p. 196.

6. *Bulletin* 187 (June 1880), pp. 311–319; 216 (November 1882), pp. 776– 786.

7. *AN* F 17 14335, Position des élèves depuis la création de l'école d'Aix jusqu'au 30 Septembre 1856.

8. *Bulletin* 46 (October 1867), p. 11; 12 (December 1932), p. 869.

9. T. Shinn and H. Paul, "The Structure of Science in France," *Contemporary French Civilization*, pp. 158–160.

10. T. Shinn, *Savoir scientifique*, p. 213.

11. Georges Ribeill did a sampling of the careers of graduates of the Ecoles d'arts et métiers during the second half of the century and obtained results similar to my own; see "La Contribution des Ingénieurs," *Culture Technique* (1984), p. 80.

12. Union des industries métallurgiques et minières, de la construction mécanique ctc. (UIMM), *Ingénieurs et cadres supérieurs*, p. 50.

13. E.N.S.A.M., *Visa pour l'avenir*, 1978, brochure; *Arts et Métiers* 7–8 (July– August 1978), pp. 21–23.

14. Fédération des associations et sociétés françaises d'ingénieurs diplômés (FASFID), *Enquête sur la situation des ingénieurs diplômés*, 1977; *Arts et Métiers* 3 (March 1978), pp. 50–51; 1 (January 1960), pp. 84–89.

15. *Arts et Métiers* 5 (May 1983), pp. 51–52.

16. Nicole Delefortrie-Soubeyroux, *Les dirigeants de l'industrie française* (Paris: Armand Colin, 1961), pp. 143, 149, 165, 205, 221, 229, 253, 259, 263. The Ecole des Mines

produced 31 graduates for the private sector (5 percent) and 71 for the public sector (7.4 percent). The Ecole supérieure d'électricité had 4 (1) private and 39 public (4 percent); other engineering schools produced 77 private (12 percent) and 173 public (18 percent); and miscellaneous, 303 (46 percent), 174 (18 percent).

17. U.I.M.M., *Ingénieurs et cadres supérieurs*, pp. 47–52.

18. Information on places of origin and work came from the database found in tables 10.9 and 10.10. The west included Brittany, the western Loire, the Vendée, and western Normandy (Calvados, Eure-et-Loire, Orne). The southwest: Acquitaine and the Midi-Pyrénées. The Center: the Auvergne, Limousin, Massif Central, and the eastern Loire (including the Yonne). The southeast: Côte-d'Azur, Languedoc, Rhône-Alpes, and Provence. The east: Alsace-Lorraine, Burgundy, and Franche-Comté. The north: Nord, Picardie, Champagne (Marne, Ardennes), and eastern Normandy (Seine-Maritime, Eure).

19. *Annuaire* 42 (1890), liste des sociétaires par départements.

20. *Bulletin* 9 (Sept. 1909), pp. 782–786.

21. *Arts et Métiers* 10 (October 1975), pp. 62–65; 4 (April 1978), p. 63; and 6 (June 1983), p. 62.

22. *Bulletin* 5 (May 1887), p. 228; 4 (April 1914), pp. 379–381. During the nineteenth century most gadzarts started out in the railways as fitters (*ajusteurs*) at 1,800 francs per year and then were allowed "to mount" the locomotive as *monteurs*, becoming firemen (*chauffeurs*), then engineers (*mécaniciens*) at 3,000 francs per year. Chief engineers were in line for promotion to roundhouse masters (*sous-chefs* and then *chefs de dépôt*), earning from 3,600 to 7,000 francs. Others became shop masters (*chefs d'atelier*) and planning supervisors (*chefs d'études*), earning middle-class salaries of 6,000 to 7,000 francs. Some became divisional or departmental heads (*chefs de section*) in the divisions of equipment or traction, and a few even reached the top positions as chief engineers and inspectors, earning 10,000 to 15,000 francs per year. From 1880 to World War I, only 8 to 10 percent reached the top posts. Although the opportunities for advancement were not as good as in industry, the railroads offered considerable benefits: heat, lodging, cheap travel, pensions, a medical plan, and unemployment insurance (by 1914).

23. *Arts et Métiers* 5 (May 1983), pp. 52–53. Gadzarts are also found in the following industries: building and public works (1,150, 7 percent); petrochemicals (1,000, 6 percent); transport (750, 4 percent); utilities (450, 2.6 percent); mines (275, 1.6 percent); food (200, 1 percent); miscellaneous (850, 5 percent); the public and military services (1,200, 7 percent).

Chapter 11

1. *Bulletin* 1 (January 1899), p. 34.

2. Jacques Payen, *Capital et machine à vapeur au XVIIIe siècle: les frères Périer et l'introduction en France de la machine à vapeur de Watt* (Paris, 1969), pp. 211, 378–383. Charles Ballot, *L'introduction du machinisme dans l'industrie française* (Paris: Rieder, 1923). Arthur L. Dunham, *The Industrial Revolution in France, 1815–1848* (New York: Exposition Press, 1955), p. 114. J. H. Clapham, *Economic Development of France and Germany, 1815–1914* (Cambridge: Cambridge University Press, 1966), p. 235, and Peter N.

Stearns, *Paths to Authority: The Middle Class and the Industrial Labor Force in France, 1820–1848* (Urbana: University of Illinois Press, 1978), p. 89.

3. James Edmonson, *The Machine Building Industry*, p. 220, also pp. 1–17. Three were gadzarts—Michel Cazalis, Jean-Jacques Meyer, and Pierre Saulnier; the others were Ernest Gouin and Emile Martin of the Ecole polytechnique and Gustave Hallet of the Ecole des arts industriels de Paris. Several others had taken courses at the Conservatoire des arts et métiers in Paris.

4. Ibid., pp. 291–296, 305.

5. Jean Vial, *Industrialisation de la sidérurgie* (Paris: Mouton, 1967), pp. 36–45. Jean-Jacques Meyer (Châlons, 1832), was an early manufacturer of locomotives in Alsace whose firm was absorbed by A. Koechlin.

6. *Recueil des machines, instruments et appareils servant à l'économie rurale et industrielle* (Paris, 1819–1852), and *Publication industrielle des machines, outils et appareils les plus perfectionnés et les plus récents*, published biannually in Paris, 32 vols., 1841–1890. Armengaud published the *Nouveau Cours raisonné de dessin industriel* (Paris, 1848), and he and his brother Charles also published *Le Génie industriel, Revue des inventions en France et à l'étranger*, Paris, 1851–1871.

7. J. E. Armengaud, "Système uniforme de filets de vis, et de proportions à adopter dans les ateliers de construction pour les boulons et les écrous," *Publication industrielle* 8 (1853), pp. 33–34.

8. Edmonson, pp. 178–216, especially p. 179. One other industrial journal, *Le Moniteur industriel*, was published at this time but was less technical in nature.

9. *Bulletin* 2 (February 1891), p. 98; 4 (April 1893).

10. J. Vial, *Industrialisation de la sidérurgie*, p. 44, and B. Gille, *Recherches*, pp. 120–23. See also Jules-François Turgan, *Les grandes usines, Etudes industrielles en France et à l'étranger*. 2nd ed., 15 vols. (Paris: Míchel Lévy Frères, 1866–1884), II. Derosne et Cail, p. 11.

11. *Bulletin* 12 (December 1912), pp. 1331–1333; and the *Enquête sur l'enseignement professionnel* (1864), Houel's observations to the commission, I, pp. 402–403.

12. Edmonson, pp. 310–316. Armengaud credited Houel with the success of Cail in manufacturing locomotives, "Fabrication de machines locomotives," *Publication industrielle* 6 (1848), p. 196.

13. A. Morin, *L'enseignement industriel*, p. 8.

14. *Bulletin* 11 (November 1900), pp. 821–822, 920–924. Among other gadzarts who became directors at the Cail Company, note Auguste Mesnard (Angers, 1841), Auguste-Adolphe Thomas (Châlons, 1871), and Louis-Emile Laurent (Châlons, 1868, and the Ecole centrale, 1876). *Bulletin* 85 (December 1871), p. 714. See also Louis Bergeron, *Les capitalistes en France, 1780–1914* (Paris: Gallimard, 1978), p. 72.

15. *Bulletin* 11 (November 1900), pp. 821–822; 12 (December 1900), pp. 920–924; 7 (July 1929), p. 759.

16. *Bulletin* 9 (September 1892), p. 556.

17. Charles F. Sabel, *Work and Politics: The Division of Labor in Industry* (Cambridge: Cambridge University Press, 1982), pp. 40–42. See also Susan Berger and Michael J. Piore, *Dualism and Discontinuity in Industrial Society* (Cambridge: Cambridge University Press, 1980).

18. Patrick O'Brien and Caglar Keyder, *Economic Growth in Britain and France 1780–1914*, pp. 174–198.

19. *Enquête sur l'enseignement professionnel* (1864), I, p. 388.

20. Ministère de l'Agriculture, du Commerce, et des Travaux publics, *Rapport et Notes*, 1865, pp. 28–35, 83–89.

21. Morin, *L'enseignement industriel*, p. 8.

22. *Annuaire* 10 (1857), pp. 207–213; 18 (1865), pp. 407–410. Among other early graduates who succeeded, note François-Xavier Jourdain (Châlons, 1812), a major manufacturer of textiles at Altkirch in Alsace, *Annuaire* 17 (1864), p. 250.

23. *L'Ingénieur des Arts et Métiers* 3 (1923), p. 32; *Livre d'Or*, p. 718.

24. *ENSAM, Ecole Nationale Supérieure d'Arts et Métiers* (Paris: Information Propagande Française, 1969), pp. 282–283.

25. Ibid., p. 284.

26. J. Clapham, *Economic Development*, p. 146.

27. A. Dunham, *Industrial Revolution*, p. 441.

28. J. Jacquet, ed., *Les Cheminots dans l'histoire sociale de la France* (Paris: Editions Sociales, 1967).

29. *Bulletin* 181 (December 1879), p. 722.

30. P. Popin, *Les Gadz'arts*, pp. 61–66; *Bulletin* 1 (January 1898), pp. 52–71.

31. *Bulletin* 41 (May 1867), pp. 24–29; 2 (February 1892), pp. 143–154. See also Jacques Boudet, *Le monde des affaires en France de 1830 à nos jours* (Paris: Société d'Edition de Dictionnaires et Encyclopédies, 1952), pp. 80–99.

32. *Bulletin* 9 (September 1882), pp. 607–612; Boudet, pp. 125–127.

33. *L'Ingénieur des Arts et Métiers* 3 (1923), p. 32.

34. *Liste générale des anciens élèves des Ecoles Nationales d'Arts et Métiers* (1900), p. 391.

35. *Bulletin* 11 (November 1900), pp. 877–887.

36. *Arts et Métiers* 1 (January 1934), p. 13; 1 (July 1960), p. 61; and 7 (July 1962), p. 55. René Sédillot, *La Maison de Wendel de mil sept cent quatre à nos jours* (Paris, 1958).

37. *Arts et Métiers* 3 (March 1961), pp. 32–40.

38. *Bulletin* 1 (January 1924), p. 54.

39. *Bulletin* 7 (July 1926), pp. 425–427; *Arts et Métiers* 3 (March 1961), p. 35.

40. *Bulletin* 1 (January 1924), p. 54; *Arts et Métiers* 11 (November 1937), pp. 703–705.

41. *Arts et Métiers* 3 (March 1961), p. 38.

42. *Bulletin* 106 (January 1930), pp. 60–63; *Arts et Métiers* 3 (March 1961), p. 38. See also Stephen Crawford, "The Work and Values of French Engineers in Traditional and Advanced Industries" (Ph.D. dissertation, Columbia University, 1985).

43. *Arts et Métiers, Revue Technique Mensuelle* 12 (December 1933), pp. 297–309.

44. Popin, *Les Gadz'arts*, pp. 119–120.

45. *Arts et Métiers* 7–8 (July–August 1952), pp. 13–16.

46. *Ingénieurs Arts et Métiers* (September 1941), p. 5; *Arts et Métiers* 10 (October 1960), p. 61.

47. *Bulletin* 12 (December 1923), p. 1110; 1 (Jan.–Feb. 1948), p. 19. See also, J. M. Laux, *In First Gear: The French Automobile Industry to 1914* (Montreal: McGill-Queens, 1976).

48. *Bulletin* 7 (July 1926), pp. 251–256; *Ingénieurs Arts et Métiers* 1–2 (January–February 1948), p. 19.

49. *ENSAM*, pp. 146–148; *Arts et Métiers* 6 (June 1962), p. 115.

50. *Arts et Métiers* 1 (January 1968), p. 39, and 8–9 (Aug.–Sept. 1973), p. 54.

51. *Arts et Métiers* 2 (February 1959), p. 77; *Livre d'Or*, p. 707.

52. *Bulletin* 12 (December 1924), pp. 927–934.

53. *Ingénieurs Arts et Métiers* 260 (December 1947), p. 4; *Arts et Métiers* 10 (October 1953), p. 19; *Livre d'Or*, p. 687.

54. *Arts et Métiers* 8 (August 1948), pp. 15–18; 3 (March 1969), p. 20. Note also J. P. Picard and Alfred Lafont, both of Aix (1919), who directed the reconstruction of the French merchant marine after World War II and later designed the ocean liner *France*, launched in 1960, a technical if not a commercial success, *Arts et Métiers* 10 (October 1963), p. 40.

55. *Arts et Métiers* 10 (October 1966), p. 29, and 5 (May 1967), p. 56.

56. Ibid. 10 (October 1966), pp. 30–33.

57. Ibid., and *Ingénieurs Arts et Métiers* 5–6 (May–June 1947), p. 5.

58. *Arts et Métiers* 5 (May 1956), p. 63. See also A. Odier's book on the heroic period of aviation in France, *Souvenirs d'une vieille tige* (Paris: Fayard, 1955).

59. *Bulletin* 12 (December 1935), p. 854; *Arts et Métiers* 10 (October 1966), pp. 27–42.

60. *Arts et Métiers* 5 (May 1967), p. 56.

61. Ibid. 10 (October 1966), pp. 33–37; *Bulletin* 12 (December 1935), p. 854.

62. *Arts et Métiers* 4 (April 1973), pp. 31–39.

63. Ibid. 10 (October 1954), p. 58; 10 (October 1966), p. 35; *Livre d'Or*, pp. 704–707.

64. *Arts et Métiers* 11 (November 1964), pp. 39–41; 10 (October 1966), pp. 34–42.

65. Ibid., and 5 (June 1981), pp. 27–29.

66. *Arts et Métiers* 10 (October 1979), p. 50.

67. *Livre d'Or*, p. 716.

68. *Ingénieurs Arts et Métiers* 3 (1923), p. 33; *Livre d'Or*, pp. 697–700.

69. *Annuaire* 2 (1849), pp. 79–82.

70. *Bulletin* 1 (January 1893), pp. 45–54, and Vial, *Le monde des affaires*, p. 395.

71. Georges Duveau, *La pensée ouvrière*, pp. 270–271.

72. *Bulletin* 2 (Feb. 1892), pp. 159–168; 10 (Oct. 1893), pp. 771–773, 787, 793; 7 (July 1896), pp. 495–521; and 11 (Nov. 1899), pp. 710–718. See also obituary notices April 1894 and May 1898, and the *Liste générale*, 1922–1923, pp. 454 and 719. Arbel was the son of a laborer, Brunon of a machine shop owner, Imbert of a blacksmith, and Savary of a worker. See also A. Robert, *Dictionnaire des Parlementaires français, 1789–1899*, 4 vols., Paris, 1891.

73. *Bulletin* 2 (February 1897), pp. 167–169; 2 (February 1899), pp. 183–191; 6 (June 1904), pp. 637–640.

74. *Arts et Métiers* 10 (October 1957), p. 95.

75. Ibid. 7 (July 1957), p. 46.

Chapter 12

1. P. Bourdieu and J. C. Passeron, *Les Héritiers* and *La Reproduction*; Michel Crozier, *The Bureaucratic Phenomenon*; Ezra Suleiman, *Elites in French Society*, pp. 276–281. In *The European Executive* (pp. 41 and 149), David Granick stated: "Whatever business talent might be found in the sons of manual workers, employees, and farmers has been disqualified from the race at the beginning. It is true that a similar disqualification occurs to some extent in all stable economies, but France is certainly outstanding." See also Nicole Delefortrie-Soubeyroux, *Les dirigeants de l'industrie française*, p. 130, who discovered a number of gadzarts in top industrial positions but concluded: "Il est vrai que parmi eux figurent nombre d'anciens élèves des Arts et Métiers qui sont en réalité des ingénieurs, mais on le constate souvent, qui débutent à des postes d'exécution subalternes. Il faut en tenir compte, car nous ne considérons pas le titre, mais la fonction professionnelle (dessinateurs ou chefs d'ateliers)."

2. Joseph Ben-David, *The Scientist's Role in Society*, pp. 88–107. Robert Gilpin, *France in the Age of the Scientific State*, pp. 77–123. R. R. Locke, *Industrial Development*, pp. 3–23, 42–57.

3. F. Artz, *Technical Education, 1500–1850*. R. Taton, *Enseignement des sciences*. T. Shinn, *Savoir scientifique*. J. Weiss, *Technological Man*.

4. J. P. Guinot, *Formation professionnelle*, pp. 149–189, 239–270. A. Léon, *Apprentissage*, pp. 19–44.

5. J. Edmonson, *Machine Building Industry*, pp. 220, 403, and 560. At the end of the century, 41 percent of the membership of the Chambre Syndicale des Mécaniciens, Chaudronniers et Fondeurs were graduates of the Ecole centrale, 19 percent of the Arts et Métiers, 7 percent of the Ecole polytechnique and its écoles d'application, and 29 percent had no formal education.

6. The Republic's claims to great achievements in primary education may have been somewhat exaggerated in view of advances in school attendance during the period before 1880, cf. Raymond Grew, Patrick Harrigan, and James Whitney, "La Scolarisation en France, 1829–1906," *Annales: Economies, Sociétés, Civilisations* 1 (January–February 1984), pp. 116–157. On the relationship of primary to higher primary and intermediate professional education in the Gard, the Ille-et-Vilaine, and the Nord, see Robert Gildea, *Education in Provincial France 1800–1914: A Study of Three Departments* (Oxford: Oxford University Press, 1983), pp. 311–313, 332–347, 369–370.

7. Michel Bouillé, *Enseignement technique*.

8. Paul Nizan, *Antoine Bloyé*.

9. M. Foucault, *Surveiller et punir*, pp. 136–196.

10. P. Gélineau, *Gadz'arts*, p. 195.

Bibliography

Archival Sources

Archives Nationales (*AN*) F 12 (Ministry of Industry and Commerce) 1084–1095. Création des Ecoles d'arts et métiers, An XII–1826, personnel, contrôle, discipline.

F 12 1096–1129^b. Ecole d'Angers, d'abord à Beaupréau. An XI–1829.

F 12 1130^a–1226^b. Ecole de Châlons, d'abord à Compiègne. An X–1841.

F 12 4867–4879. Ecoles d'arts et métiers: Aix, Angers, Châlons: personnel, inspection, discipline, 1806–1881.

F 12 5775–5782. Ecoles d'arts et métiers, personnel, 1820–1850.

F 17 (L'Instruction Publique) 14317–14326. Ecoles nationales d'arts et métiers, 1814–1923: origins, inspection, programs, alumni association, World War I.

F 17 14327–31. Ecole nationale d'arts et métiers de Châlons. An XII–1917.

F 17 14332–34. Ecole nationale d'arts et métiers de Beaupréau puis d'Angers. An XII–1923.

F 17 14335–36. Ecole nationale d'arts et métiers d'Aix, 1837–1924.

F 17 14337–40. Ecole nationale d'arts et métiers de Lille, 1879–1927.

F 17 14341–44. Ecole nationale d'arts et métiers de Cluny, 1891–1932.

F 17 14345–47. Ecole nationale d'arts et métiers de Paris, 1901–1926.

F 17 14348–64. Ecoles nationales professionnelles, creation, personnel, programs, recruitment, placement, scholarships.

F 17 8701–8731. L'enseignement secondaire spécial, 1847–1890.

F 17 8732–8752. L'école normale secondaire spéciale de Cluny, 1866–1891.

F 17 9813–9819. Renseignements statistiques sur les élèves sortis des écoles primaires supérieures, 1887–1890.

F 17 9820–9824. Les écoles primaires supérieures, 1830s–1840s.

F17 11706–11708. L'enseignement professionnel, 1850s–1880s.

Statistical Studies and Official Publications

Enquêtes de l'Union des industries métallurgiques et minières sur la situation des ingénieurs et cadres supérieurs et sur la prévision et les besoins dans les industries des métaux, 1956, 1962, 1970, and 1977.

L'Expansion-SOFRES: *Les salaires des cadres.* Done each year since 1970.

Fédération des associations et sociétés françaises d'ingénieurs diplômés. *Enquêtes socio-économiques sur la situation des ingénieurs diplômés.* "FASFID" published seven major socioeconomic studies on certified engineers in 1958, 1963, 1967, 1971, 1974, 1977, and 1983.

Institut national de la statistique et des études économiques (INSEE). *Enquête formation-qualification professionnelle de 1970* (on executives generally, including engineers).

Institut national de la statistique et des études économiques (INSEE). *Tableaux de l'Économie française.* Paris: INSEE, 1978.

Ministère de l'Agriculture, du Commerce, et des Travaux Publics. *Enquête sur l'enseignement professionnel ou recueil de dépositions faites en 1863 et 1864 devant la commission de l'enseignement professionnel.* 2 vols. Paris: Imprimerie Impériale, 1864–1865.

Ministère de l'Agriculture, du Commerce, et des Travaux Publics. Commission de l'enseignement technique, by A. Morin. *Rapport et Notes.* Paris: Imprimerie Impériale, 1865.

Ministère du Commerce, de l'Industric, etc. Conseil supérieur de l'enseignement technique. *Rapport sur l'organisation de l'enseignement technique,* by H. Tresca. Paris: Imprimerie nationale, 1885.

Ministère du Commerce, de l'Industrie, des Postes et des Télégraphes. Direction de l'enseignement technique. *L'Enseignement technique en France, Etude publiée à l'occasion de l'exposition de 1900.* 5 vols. Paris: Imprimerie nationale, 1900.
Vol. 1: *Etablissements nationaux et Ecoles supérieures de commerce,* by Paul Buquet.
Vols. 2–3: *Ecoles pratiques de commerce et d'industrie,* by Félix Martel.
Vols. 4–5: *Etablissements divers* by MM. Couriot and Delmas.

Ministère du Commerce, de l'Industrie, des Postes et des Télégraphes. Conseil supérieur de l'enseignement technique. *Rapport sur la situation de l'enseignement technique en France en 1904, présenté par Cohendy: procès-verbaux des séances.* Paris: Imprimerie nationale, 1905.

Ministère du Commerce, de l'Industrie, des Postes et des Télégraphes. Conseil supérieur du Travail. *Enquête récente sur l'enseignement professionnel en France, Rapport de M. Briat au nom de la commission permanente: procès-verbaux des séances.* Paris: Imprimerie nationale, 1905.

Union des industries métallurgiques. *Le personnel dans l'industrie des métaux.* Paris: U.I.M.M., 1963.

Union des industries métallurgiques et minières, de la construction mécanique, électrique et métallique et des industries qui s'y rattachent. *Ingénieurs et cadres supérieurs, situation actuelle et prévision des besoins dans les industries des métaux.* Paris: U.I.M.M., 1956.

Periodicals

Annuaire de la Société des anciens élèves des Ecoles d'Arts et Métiers (Annuaire). 1848   . See the *Liste générale alphabétique et par promotions des anciens élèves des Ecoles nationales d'Arts et Métiers.* Paris: A + M, 1900 and 1923.

Arts et Métiers, Revue mensuelle. 1951—.

Arts et Métiers, Revue technique mensuelle. 1930–1940.

Bulletin administratif (mensuel) de la Société des anciens élèves des Ecoles d'Arts et Métiers (Bulletin). 1861–1937.

Bulletin de l'enseignement technique. Published monthly by the Ministry of Industry and Commerce, 1898–1914.

Bulletin mensuel de l'Association amicale des anciens élèves de l'Ecole normale secondaire spéciale de Cluny.

Bulletin mensuel de la Société amicale des anciens élèves des Ecoles nationales professionnelles 1920–1950.

Culture technique, Revue trimestrielle. Edited by the Centre de Recherche sur la Culture Technique (CRCT). 1979—.

L'Education professionnelle. Published by the Fédération nationale des anciens élèves des enseignements techniques et professionnels.

L'Enseignement technique, Revue trimestrielle. Published by the Association française pour le développement de l'enseignement technique (AFDET). 1954—.

La Formation professionnelle. Published quarterly by the AFDET. 1904–1914, 1919–1939.

Le Génie industriel, Revue des inventions en France et à l'étranger. 1851–1871.

Ingénieurs Arts et Métiers. Monthly, 1938–1950.

Manuel général de l'Instruction publique. Published by the Ministry of Public Instruction.

Le Monde de l'Education. Monthly, since 1974.

Publication industrielle des machines, outils et appareils les plus perfectionnés et les plus récents. 32 vols. 1841–1890.

Recueil des machines, instruments et appareils servant à l'économie rurale et industrielle. 1819–1852.

Revue de l'enseignement secondaire spécial et de l'enseignement professionnel. Monthly, 1878 to the end of the century.

Revue de l'enseignement technique. Published monthly by the AFDET, 1910–1914.

Le Technicien. Published monthly by the alumni association of the Ecoles nationales professionnelles, 1950–1961.

Technique, Art, Science (TAS). Published monthly by the AFDET and Ministry of National Education, 1946–1977.

Published Works

Alain (Emile Chartier). *Propos sur L'Education.* Paris: Presses Universitaires de France, 1963.

Anderson, Robert. *Education in France, 1848–1870.* Oxford: Oxford University Press, 1975.

Armitage, W. H. G. *The Rise of the Technocrats.* London: Routledge and Kegan Paul, 1965.

Artz, Frederick. *The Development of Technical Education in France 1500–1850.* Cambridge, Mass.: The MIT Press, 1966.

Astier, Placide, and Cuminal, I. *L'Enseignement technique, industriel et commercial en France et à l'étranger*. 2nd ed. Paris, 1912.

Ballot, Charles. *L'introduction du machinisme dans l'industrie française*. Paris: Rieder, 1923.

Baudant, Alain. *Pont-à-Mousson, 1918–1939: Stratégies industrielles d'une dynastie lorraine*. Paris: Publications de la Sorbonne, 1980.

Ben-David, Joseph. *The Scientist's Role in Society: A Comparative Study*. Englewood Cliffs, N.J.: Prentice-Hall, 1971.

Berger, Susan, and Piore, Michael J. *Dualism and Discontinuity in Industrial Society*. Cambridge, England: Cambridge University Press, 1980.

Bergeron, Louis. *Les capitalistes en France 1780–1914*. Paris: Gallimard, 1978.

Boltanski, Luc. *Les cadres: la formation d'un groupe sociale*. Paris: Les Editions de Minuit, 1982.

Boudet, Jacques. *Le monde des affaires en France de 1830 à nos jours*. Paris: Société d'Edition de Dictionnaires et Encyclopédies, 1952.

Bouillé, Michel. *Enseignement technique et idéologies au XIXe siècle*. Thesis, Ecole pratique des Hautes Etudes, 1972.

Bourdieu, Pierre. *La Distinction*. Paris: Editions de Minuit, 1979.

Bourdieu, Pierre, and Passeron, J. C. *Les Héritiers*. Paris: Les Editions de Minuit, 1964.

Bourdieu, Pierre, and Passeron, J. C. *La Reproduction, éléments pour une théorie du système d'enseignement*. Paris: Les Editions de Minuit, 1970.

Briand, J. P., Chapoulie, J. M., and Peretz, H. "Les conditions institutionnelles de la scolarisation secondaire des garçons entre 1920 et 1940." *Revue d'histoire moderne et contemporaine* XXV (July–Sept. 1979), pp. 391–421.

Briand, J. P., Chapoulie, J. M., and Peretz, H. "Les statistiques scolaires comme représentation et comme activité." *Revue française de sociologie* XX (1979), pp. 669–702.

Buisson, Ferdinand. *L'Enseignement primaire supérieur et professionnel*. Paris, 1887.

Calvert, Monte. *The Mechanical Engineer in America, 1830–1910, Professional Cultures in Conflict*. Baltimore: Johns Hopkins University Press, 1967.

Cameron, Rondo. *France and the Economic Development of Europe, 1800–1914*. Princeton, N.J.: Princeton University Press, 1961.

Chandler, Alfred D. *The Visible Hand: The Managerial Revolution in American Business*. Cambridge, Mass.: Harvard University Press, 1977.

Chandler, Alfred D., and Daems, Herman, eds. *Managerial Hierarchies: Comparative Perspectives on the Rise of Modern Industrial Enterprise*. Cambridge, Mass.: Harvard University Press, 1980.

Chevalier, Michel. *Des intérêts matériels en France: Travaux publics, routes, canaux, chemins de fer*. Paris, 1839.

Clapham, J. H. *Economic Development of France and Germany 1815–1914*. 4th ed. Cambridge: Cambridge University Press, 1966.

Cogniot, Georges, ed. *Le Plan Langevin-Wallon, La nationalisation de l'enseignement*. Paris, 1962.

Collins, Randall. *The Credential Society: An Historical Sociology of Education and Stratification.* New York: Academic/Harcourt Brace Jovanovich, 1979.

Corbon, Anthime. *L'enseignement professionnel.* Paris, 1859.

Crawford, Stephen. "The Work and Values of French Engineers in Traditional and Advanced Industries." Ph.D. dissertation, Columbia University, 1985.

Crouzet, François, "An Annual Index of French Industrial Production in the Nineteenth Century." In *Essays in French Economic History*, pp. 245–278. Edited by Rondo Cameron. Homewood, Ill.: Irwin, 1970.

Crozier, Michel. *The Bureaucratic Phenomenon.* Chicago: University of Chicago Press, 1964.

Dallay, Guy. *Qu'est-ce qu'un Gadz'arts?* Talence: the author, 1943.

Day, C. R. "Technical and Professional Education in France: The Rise and Fall of L'Enseignement Secondaire Spécial, 1865–1902." *Journal of Social History* 6 (1972–1973), pp. 177–201.

Day, C. R. "Education, Technology and Social Change in France: The Short Unhappy Life of the Cluny School, 1866–1891." *French Historical Studies* 7 (1974), pp. 427–444.

Day, C. R. "The Making of Mechanical Engineers in France: The Ecoles d'Arts et Métiers, 1803–1914." *French Historical Studies* 10 (1978), pp. 439–460.

Day, C. R. "Education for the Industrial World: Technical and Modern Instruction in France Under the Third Republic, 1870–1914." In *The Organization of Science and Technology in France, 1808–1914*, pp. 127–153. Edited by R. Fox and G. Weisz. Cambridge: Cambridge University Press, 1980.

De La Rochefoucauld, Jean-Dominique, Wolikow, Claudine, and Ikni, G. *Le Duc de La Rochefoucauld-Liancourt de Louis XV à Charles X, un grand seigneur patriote et le mouvement populaire.* Paris: Perrin, 1981.

Delefortrie-Soubeyroux, Nicole. *Les dirigeants de l'industrie française.* Paris: Armand Colin, 1961.

Dunham, Arthur Louis. *The Industrial Revolution in France, 1815–1848.* New York: Exposition Press, 1955.

Dupin, Charles. *Avantages sociaux d'un enseignement public appliqué à l'industrie.* Paris: Bachelier, 1824.

Durkheim, Emile. *L'évolution pédagogique en France.* Paris: Alcan, 1938.

Duveau, Georges. *La pensée ouvrière sur l'éducation pendant la Seconde République et le Second Empire.* Paris: Domat, 1948.

Edmonson, James M. *From Mécanicien to Engineer, Technical Education and the Machine-Building Industry in Nineteenth-Century France.* New York: Garland, 1986.

Euvrard, F. *Historique de l'Ecole nationale d'arts et métiers de Châlons-sur-Marne, 1780–1895.* Châlons-sur-Marne: Union Républicaine, 1895.

Ferdinand-Dreyfus, Camille. *Un philanthrope d'autrefois, La Rochefoucauld-Liancourt, 1747–1827.* Paris: Plon, 1903.

Ferraud, Georges, and Martel, Félix. "Ecoles primaires supérieures, écoles d'apprentissage, et écoles nationales professionnelles." *Mémoires et documents scolaires*. 2nd series, no. 9. Paris: Musée Pédagogique, 1889.

Foucault, Michel. *Surveiller et punir: naissance de la prison*. Paris: Gallimard, 1975.

Fourastié, Jean. *Le grand espoir du XXe siècle; progrès technique, progrès économique, progrès social*. Preface by André Siegfried. Paris: Presses Universitaires de France, 1949.

Fox, Robert W. "Education for a New Age: the Conservatoire des Arts et Métiers, 1815–1830." In *Artisan to Graduate*. Edited by D. S. L. Cardwell. Manchester: Manchester University Press, 1974.

Fox, Robert, and Weisz, George, eds. *The Organization of Science and Technology in France, 1808–1914*. Cambridge: Cambridge University Press, 1980.

Friedmann, Georges. *Industrial Society*. Edited and with an introduction by Harold L. Shepherd. Glencoe, Ill.: Free Press, 1955.

Gélineau, Paul. *Gadz'arts*. Paris: the author, 1910.

Gerbod, Paul. *La condition universitaire en France au XIXe siècle*. Paris: P.U.F., 1965.

Gerstl, J. E., and Hutton, S. P. *Engineers: The Anatomy of a Profession—A Study of Mechanical Engineers in Britain*. London: Tavistock, 1966.

Gerth, H. H., and Mills, C. W., eds. and trans. *From Max Weber: Essays in Sociology*. New York: Galaxy, 1958.

Gildea, Robert. *Education in Provincial France 1800–1914: A Study of Three Departments*. Oxford: Oxford University Press, 1983.

Gille, Bertrand. *Recherches sur la formation de la grande entreprise capitaliste, 1814–1848*. Paris: S.E.V.P.E.N., 1959.

Gille, Bertrand. *Le sidérurgie française au XIX siècle*. Geneva: Librairie Droz, 1968.

Gilpin, Robert. *France in the Age of the Scientific State*. Princeton, N.J.: Princeton University Press, 1968.

Granick, David. *The European Executive*. N.Y.: Doubleday, 1962.

Gréard, Octave. *Education et Instruction*. 2 vols. Paris, 1889.

Grelon, André. "L'Education des cadres: la question des aspirations professionnelles chez les futurs cadres d'entreprise." Thèse de troisième cycle, Université de Paris VII, 1983.

Grew, Raymond, Harrigan, Patrick, and Whitney, James. "La Scolarisation en France, 1829–1906." *Annales: Economies, Sociétés, Civilisations* 1 (January–February 1984), pp. 116–157.

Guerlac, Henry. "Science and French National Strength." In *Modern France, Problems in the Third and Fourth Republics*, pp. 81–105. Edited by Edward Mead Earle. Princeton, N.J.: Princeton University Press, 1951.

Guettier, André. *Histoire des Ecoles nationales d'arts et métiers*. 2nd ed. Paris: Dejay, 1880.

Guillou, Jules, et al. *Un siècle sous les cloîtres du Ronceray, Ecole nationale des arts et métiers d'Angers*. Angers: the authors, 1956.

Guinot, Jean-Pierre. *Formation professionnelle et travailleurs qualifiés depuis 1789*. Paris: Domat, 1946.

Haby, René. *Propositions pour une modernisation du système éducatif*. Paris: La Documentation française, 1975.

Harrigan, Patrick J. *Mobility, Elites, and Education in French Society of the Second Empire.* Waterloo, Ontario: Wilfred Laurier University Press, 1980.

Horvath-Peterson, Sandra. *Victor Duruy.* Baton Rouge: Louisiana State University Press, 1984.

Jacquet, J., ed. *Les cheminots dans l'histoire sociale de la France.* Paris: Editions Sociales, 1967.

Kindleberger, Charles P. "Technical Education and the French Entrepreneur." In *Enterprise and Entrepreneurs in Nineteenth- and Twentieth-Century France*, pp. 3–39. Edited by E. C. Carter, R. Forster, and J. N. Moody. Baltimore: Johns Hopkins University Press, 1976.

Labarde, H. *Le Ronceray, Ecole de Gadz'arts, 1527–1927.* Pau: the author, 1941.

Labbé, Edmond. *Les outils de perçage, leçon type de technologie, faite à des élèves de cours professionnels.* Paris, 1926.

Langevin, Paul. *La pensée et l'action, textes recueillis.* Paris, 1950.

Laux, J. M. *In First Gear: The French Automobile Industry to 1914.* Montreal: McGill-Queens, 1976.

Layton, Edwin T., Jr. *The Revolt of the Engineers: Social Responsibility and the American Engineering Profession.* Cleveland: Case Western Reserve University Press, 1971.

Leblanc, René. *La réforme des Ecoles primaires supérieures.* Paris, 1907.

Le Chartier, Eugène. *La France et son Parlement, annuaire des électeurs et des parlementaires de 1871 à 1912.* Paris, 1912.

Léon, Antoine. *Formation générale et apprentissage du métier.* Paris: Presses Universitaires de France, 1965.

Le Play, Frédéric. *Les ouvriers européens, Etude sur les travaux, la vie domestique et la condition des populations ouvrières de l'Europe.* Paris, 1855.

Leroy, L. M. *L'éducation nationale au XXe siècle.* Paris, 1914.

Leroy, L. M. *Vers l'éducation nouvelle.* Paris, 1906.

Levasseur, Emile. *L'instruction primaire et professionnelle en France sous la Troisième République.* Paris, 1906.

Lévy-Leboyer, Maurice. "Innovation and Business Strategies in Nineteenth- and Twentieth-Century France." In *Enterprise and Entrepreneurs*, pp. 87–135. Edited by E. C. Carter, R. Foster, and J. N. Moody. Baltimore: Johns Hopkins University Press, 1976.

Locke, Robert R. *Industrial Development and the Social Fabric: The End of the Practical Man: Entrepreneurs and Higher Education in Germany, France and Great Britain, 1880–1940.* Greenwich, Conn.: J.A.I. Press, 1984.

Luc, Hippolyte. "Les problèmes actuels de l'Enseignement technique." *L'Enseignement technique* 1 (1938), pp. 8–35.

Lundgreen, Peter. *Bildung und Wirtschaftswachstum im Industrialisierungsprozess des 19. Jahrhunderts.* Berlin: Colloquium, 1973.

Lundgreen, Peter. "Industrialization and the Educational Formation of Manpower in Germany." *Journal of Social History* 9 (1975–1976) pp. 64–80.

Lundgreen, Peter. "Natur- und Technikwissenschaften an deutschen Hochschulen,

1870–1970: Einige quantitative Entwicklungen." In *Wissenschaft und Gesellschaft, Beiträge zur Geschichte der Technischen Universität Berlin, 1879–1979*. Berlin: Springer-Verlag, 1979.

Lundgreen, Peter. "The Organization of Science and Technology in France: A German Perspective." In *The Organization of Science and Technology in France, 1808–1914*. R. Fox and G. Weisz, eds. Cambridge: Cambridge University Press, 1980.

Mallet, Serge. *La nouvelle classe ouvrière*. Paris: Editions du Seuil, 1963. Second edition, 1969.

Markovitch, T. J. *L'Industrie française de 1789 à 1964: Conclusions générales*. vol. 7 in *Histoire quantitative de l'économie française*. Paris: Institut de science économique appliquée, 1966.

Metton, A. *Les Gadz'arts, Les Ingénieurs des Arts et Métiers*. Paris, 1925.

Moody, Joseph N. *French Education Since Napoleon*. Syracuse, N.Y.: Syracuse University Press, 1978.

Morin, Arthur, and Tresca, Henri. *De l'organisation de l'enseignement industriel et de l'enseignement professionnel*. Paris, 1862.

Moutet, Aimée. "Ingénieurs et rationalisation dans l'industrie française de la Grande Guerre au Front Populaire." *Culture Technique* 12 (March 1984), pp. 137–153.

Nizan, Paul. *Antoine Bloyé*. New York: Monthly Review Press, 1973.

Nye, Mary Jo. *Science in the Provinces, Scientific communities in France, 1860–1930*. Berkeley: University of California, 1986.

Odier, Antoine. *Souvenirs d'une vieille tige*. Paris: Fayard, 1955.

O'Brien, Patrick K., and Keyder, Caglar. *Economic Growth in Britain and France, 1780–1914: Two Paths to the Twentieth Century*. London and Boston: Allen and Unwin, 1978.

Paul, Harry W. *The Sorcerer's Apprentice, the French Scientist's Image of German Science, 1840–1919*. Gainsville: University of Florida Press, 1972.

Paul, Harry W. "Apollo Courts the Vulcans: The Applied Science Institutes in Nineteenth-Century Science Faculties." In *The Organization of Science and Technology in France, 1808–1914*, pp. 155–181. Edited by R. Fox and G. Weisz. Cambridge: Cambridge University Press, 1980.

Paul, Harry, W. "The Issue of Decline in Nineteenth-Century French Science." *French Historical Studies* 7 (1972), pp. 416–450.

Payen, Jacques. *Capital et machine à vapeur au XVIIIᵉ siècle: les frères Périer et l'introduction en France de la machine à vapeur de Watt*. Paris, 1969.

Payen, Jacques. "Technologie de l'énergie vapeur en France dans la première moitié du XIXᵉ siècle." Thèse du 3ᵉ cycle. 3 vols. Université de Paris, 1977.

Perrucci, Robert, and Gerstl, Joel E., eds. *The Engineers and the Social System*. New York: Wiley, 1969.

Pitts, Jesse. "Change in Bourgeois France." Stanley Hoffmann et al., *In Search of France*. New York: Harper Torchbooks, 1963.

Pollard, Sidney. *The Genesis of Modern Management: A Study of the Industrial Revolution in Great Britain*. Cambridge, Mass.: Harvard University Press, 1965.

Ponteil, Félix. *Histoire de l'enseignement en France, les grandes étapes, 1789–1965*. Paris: Sirey, 1966.

Popin, Paul. *Les Gadz'arts*. Saint-Dizier: Ets. Bruillard, 1947.

Poulot, D. *Le Sublime, ou le travailleur comme il est en 1870 et ce qu'il peut être*. Paris, 1872. Reissued in 1982 by Editions Maspéro, Alain Cottereau, editor.

Poulot, D. *Manifeste d'un Bourgeois démocrate*. Paris, 1871.

Prévot, André. *L'enseignement technique chez les Frères des Ecoles Chrétiennes au XVIIIe et XIXe siècles*. Paris: Ligel, 1932.

Primault, Jean, et al. *Le Livre d'Or du Bicentenaire des Ecoles d'Arts et Métiers*. Paris: Arts et Métiers, 1980.

Prost, Antoine. *Histoire de l'enseignement en France, 1800–1967*. Paris: Armand Colin, 1968.

Redfern, W. D. *Paul Nizan, Committed Literature in a Conspiratorial World*. Princeton, N.J.: Princeton University Press, 1972.

Ribeill, Georges. "Entreprendre hier et aujourd'hui: la contribution des ingénieurs." *Culture Technique, Les Ingénieurs* 12 (March 1984), pp. 55–154.

Ringer, Fritz K. *Education and Society in Modern Europe*. Bloomington: Indiana University Press, 1979.

Roux, F. *Histoire des six premières années de l'Ecole normale spéciale de Cluny*. Alais, 1889.

Sabel, Charles F. *Work and Politics: The Division of Labor in Industry*. Cambridge: Cambridge University Press, 1982.

Sauvy, A. *La montée des jeunes*. Paris: Calmann-Lévy, 1959.

Shinn, Terry. *Savoir scientifique et pouvoir social, l'Ecole Polytechnique, 1794–1914*. Paris: Presses de la fondation nationale des sciences politiques, 1980.

Shinn, Terry. "Des Corps de l'Etat ou secteur industriel: genèse de la profession d'ingénieur, 1750–1920." *Revue française de sociologie* 19 (1978), pp. 39–71.

Shinn, Terry. "From 'corps' to 'profession': The emergence and definition of industrial engineering in modern France." In *The Organization of Science and Technology in France, 1808–1914*, pp. 183–210. Edited by R. Fox and G. Weisz. Cambridge: Cambridge University Press, 1980.

Shinn, Terry. "The French Science Faculty System 1808–1914: Institutional Change and Research Potential in Mathematics and the Physical Sciences." In *Historical Studies in the Physical Sciences*, pp. 271–332. Edited by R. McCormmach, L. Pyenson, and R. S. Turner. Baltimore: the Johns Hopkins University Press, 1979.

Sinclair, Bruce. *Philadelphia's Philosopher Mechanics: A History of the Franklin Institute, 1824–1865*. Baltimore: The Johns Hopkins University Press, 1974.

(La) Société des Ingénieurs civils de France. *125 ans de progrès technique vus à travers la Société des Ingénieurs civils de France*. Paris: ICF, 1973.

Sofer, C. *Men in Mid-Careers: A Study of British Managers and Technical Specialists*. Cambridge: Cambridge University Press, 1970.

Stearns, Peter N. *Paths to Authority: The Middle Class and the Industrial Labor Force in France, 1820–1848*. Urbana: University of Illinois Press, 1978.

Suleiman, Ezra N. *Elites in French Society: The Politics of Survival*. Princeton, N.J.: Princeton University Press, 1978.

Talbot, John. *The Politics of Educational Reform in France, 1918–1949.* Princeton, N.J.: Princeton University Press, 1969.

Taton, René. *Enseignement et diffusion des sciences en France au XVIIIᵉ siècle.* Paris: Herman, 1964.

———. *L'oeuvre scientifique de Monge.* Paris: P.U.F., 1951.

Thépot, André, ed. *L'ingénieur dans la société française.* Paris: Editions ouvrières, 1985.

Turgan, Julien-François. *Les grandes usines, études industrielles en France et à l'étranger.* 2nd ed. 15 vols. Paris: Michel Lévy frères, 1866–1884.

Vial, Jean. *L'industrialisation de la sidérurgie, 1814–1864.* Paris: Mouton, 1967.

Wallon, Henri. *De l'acte à la pensée.* Paris, 1942.

Weber, Max. *The Theory of Social and Economic Organization.* New York: Free Press, 1965.

Weiss, John H. *The Making of Technological Man: The Social Origins of French Engineering Education.* Cambridge, Mass.: The MIT Press, 1982.

Weiss, John H. "Bridges and Barriers: Narrowing Access and Changing Structure in the French Engineering Profession, 1800–1850." In *The Professions and the French State.* Edited by Gerald Geison. Philadelphia: University of Pennsylvania Press, 1984.

Weiss, John H. *Careers and Comrades: Professional Lives in a French Administration, 1804–1930.* Unpublished manuscript.

Weisz, George. *The Emergence of Modren Universities in France, 1863–1914.* Princeton, N.J.: Princeton University Press, 1983.

Wylie, Laurence. *Village in the Vaucluse.* Cambridge, Mass.: Harvard University Press, 1974.

Zeldin, Theodore. *France 1848–1955: Ambition, Love, and Politics.* London: Oxford University Press, 1973.

Index